Experimental study of the zonal-flow
in the magnetised plasmas of the stellarator experiment TJ-K

Experimental study of the zonal-flow dynamics in the magnetised plasmas of the stellarator experiment TJ-K

Von der Fakultät Energie-, Verfahrens- und Biotechnik
der Universität Stuttgart
zur Erlangung der Würde eines Doktors der
Naturwissenschaften (Dr. rer. nat) genehmigte Abhandlung

vorgelegt von

Bernhard Schmid

aus Neu-Ulm

| Hauptberichter: | Prof. Dr. U. Stroth |
| Mitberichter: | Prof. Dr. J. Starflinger |

Tag der mündlichen Prüfung: 13.04.2018

Institut für Grenzflächenverfahrenstechnik und Plasmatechnologie der
Universität Stuttgart

2018

Bibliografische Information der Deutschen Nationalbibliothek
Die Deutsche Nationalbibliothek verzeichnet diese Publikation in der
Deutschen Nationalbibliografie; detaillierte bibliografische Daten
sind im Internet über http://dnb.d-nb.de abrufbar.
1. Aufl. - Göttingen: Cuvillier, 2019
 Zugl.: Stuttgart, Univ., Diss., 2018

© CUVILLIER VERLAG, Göttingen 2019
 Nonnenstieg 8, 37075 Göttingen
 Telefon: 0551-54724-0
 Telefax: 0551-54724-21
 www.cuvillier.de

 ISBN 978-3-7369-9993-0
 eISBN 978-3-7369-8993-1

Zusammenfassung

Die Einschlussqualität von Plasmen in toroidalen Magnetfeldern wird maßgeblich durch den turbulenten Transport senkrecht zum Magnetfeld limitiert. Zonalströmungen sind dabei für die Fusionsforschung von großer Bedeutung, da vermutet wird, dass sie mit der Bildung von Transportbarrieren in Zusammenhang stehen. Diese mesoskopischen Scherströmungen tragen auf Grund ihrer Symmetrie nicht zum turbulenten Transport bei und können durch Verscherung von Wirbeln den radialen Transport unterdrücken. Dabei werden Zonalströmungen in einem Selbstorganisationsprozess von der umgebenden Plasmaturbulenz generiert indem Wirbel durch die Zonalströmung verkippt werden, was die Scherströmung weiter antreibt. Ein Maß für die Verkippung ist der sogenannte Reynolds-Stress, wobei der radiale Gradient des Flussflächenmittels die Antriebskraft der Zonalströmung darstellt. Die Dynamik gleicht dabei einer Räuber-Beute-Beziehung, bei der die Driftwellen die Beute für die Zonalströmungen sind. In Fusionsexperimenten konnte beim spontanen Übergang in ein verbessertes Einschlussregime (H-Mode) ein verstärktes Auftreten von Zonalströmungen mit den charakteristischen Räuber-Beute-Oszillationen nachgewiesen werden. Die Rolle und die genaue Wirkungsweise der Zonalströmungen bei dieser Bifurkation des Plasmaeinschlusses sind jedoch ungeklärt. Ein tieferes Verständnis der Physik der Zonalströmungen, speziell in komplexen Magnetfeldgeometrien, ist daher wünschenswert.

Diese Arbeit beschäftigt sich vorwiegend mit der Untersuchung des Antriebsmechanismus von Zonalströmungen, im Speziellen mit der Abhängigkeit von der Magnetfeldgeometrie und dem Einfluss der Kollisionalität. Dazu wurden Messungen am Stellarator-Experiment TJ-K durchgeführt, in Plasmen, die dimensional ähnlich zu Randplasmen von Fusionsexperimenten sind. Die relativ geringen Plasmatemperaturen erlauben den Einsatz von Langmuir-Sonden im gesamten Einschlussgebiet. Mit einem poloidalen Sonden Array, bestehend aus 128 Sonden mit je 32 Sonden auf vier benachbarten Flussflächen, können Dichte- und Potentialfluktuationen mit hoher räumlicher und zeitlicher Auflösung gleichzeitig über den gesamten poloidalen Umfang aufgenommen werden. Daraus lassen sich Geschwindigkeitsfluktuationen und der Reynolds-Stress, sowie dessen radiale Ableitung, direkt bestimmen. Dies bietet die Möglichkeit den Antrieb der Zonalströmungen durch Reynolds-Stress auf den Einfluss der Magnetfeldparameter hin zu analysieren. Für die Untersuchung der Stoßratenabhängigkeit des Antriebs wurde eine Vielzahl an Messungen bei unterschiedlicher Kollisionalität durchgeführt, was die Kopplung zwischen Dichte und Potential maßgeblich beeinflusst. Über die Variation der Ionenmasse von Wasserstoff bis Krypton, des Drucks und der Heizleistung kann die Kollisionalität C über vier Größenordnungen kontinuierlich vom adiabatischen Regime ($C \ll 1$) ins hydrodynamische Regime ($C \gg 1$) verändert werden.

In dieser Arbeit wurde gezeigt, dass Zonalströmungen gleichverteilt als positive und negative homogene Potentialstörungen auf einer gesamten Flussfläche mit schmaler radialer Ausdehnung auftreten, wobei der Großteil der spektralen Leistung auf Frequenzen unter 8 kHz begrenzt ist. Ein teilweise auftretender Beitrag bei höheren Frequenzen ist gering, könnte aber auf die Existenz einer geodätisch akustischen Mode hindeuten.

Der Reynolds-Stress ist nicht räumlich homogen verteilt, sondern ist besonders stark in Bereichen negativer normalen Feldlinienkrümmung κ_n und positiver geodätischer Feldlinienkrümmung κ_g. Dies gleicht der Verteilung des turbulenten Transports der eine ähnliche konzeptuelle Form aufweist. Zusätzlich hat die magnetische Verscherung (integrierte und lokale Verscherung) einen Einfluss auf die Verkippung der Wirbel und somit auf den Reynolds-Stress. Quantitativ wurde gezeigt, dass der Reynolds-Stress ausreicht um die Zonalströmung anzutreiben. Die 3-Wellenkopplung ist stark während dem Auftreten der Zonalströmung und der direkte Energietransfer kleiner Skalen in die Zonalströmung bestätigt das Bild des nichtlokalen Antriebs im k-Raum.

Die Effektivität des Antriebsmechanismus hängt von der Dichte-Potential-Kopplung und damit in erster Linie von der Kollisionalität ab. Durch die Analyse von Reynolds-Stress und dichtebasiertem Pseudo-Reynolds-Stress zeigt sich, dass bei geringer Kollisionalität die Kopplung zwischen Dichte und Potential zunimmt. Dabei steigen sowohl der Energietransfer in die Zonalströmung als auch die relative Zonalströmungsleistung stark an. Ein Anteil der Zonalströmung an der Gesamtleistung der Turbulenz von bis zu 29 % wird erreicht. Damit bestätigen die Messungen einen grundlegenden Mechanismus der Plasmaturbulenz und belegen die Bedeutung der Kollisionalität für die Entstehung großskaliger Strukturen in toroidal eingeschlossenen Plasmen.

Abstract

The confinement quality of plasmas in toroidal magnetic fields is mainly limited by the turbulent transport perpendicular to the magnetic field. Zonal flows play an important role in fusion research as they are thought to be connected to the formation of a transport barrier in the edge of the confined plasma. Because of their symmetry, these mesoscale turbulent structures do not contribute to turbulent cross-field transport and can suppress radial transport by shearing off turbulent eddies. Like in a self-organisation process, the zonal flows are generated by the ambient turbulence itself as turbulent eddies are tilted by the shear flow. For tilted vortices the so-called Reynolds stress is non-zero and the radial gradient of this flux surface averaged quantity drives the zonal flow. The dynamics resemble a predator-prey relationship, where the drift waves are the prey for the zonal flow. At the spontaneous transition to an improved confinement regime (H-mode), an increased zonal flow occurrence, with the characteristic predator-prey oscillation, was indeed confirmed by many experiments. But the mechanism behind the occurrence of this bifurcation in the plasma confinement is still not fully understood. Therefore, a deeper understanding of the zonal flow physics, especially in complex magnetic field geometries, is highly desirable.

This work concentrates on the investigation of the Reynolds stress drive of zonal flows with its connection to the geometry of the confining magnetic field and the influence of the collisionality. The measurements for this work have been conducted at the stellarator experiment TJ-K in plasmas dimensionally similar to fusion edge plasmas. The low temperatures allow the use of Langmuir probes in the entire confinement region. With a poloidal probe array, consisting of 128 Langmuir probes with 32 probes on each of four neighbouring magnetic flux surfaces, density and potential fluctuations can be acquired with high spatial and temporal resolution on the complete poloidal circumference. Thus, velocity fluctuations as well as the turbulent Reynolds stress, and its gradient, are available. This gives the possibility to study the Reynolds stress drive of zonal flows with respect to the influence of the magnetic field parameters. For the investigation of the collisional dependence of the driving mechanism, a multitude of measurements at different collisionality, which determines the coupling strength between density and potential, has been performed. By changing the ion mass from hydrogen to krypton, the pressure, and the heating power, the collisionality C can be continuously varied by about four orders of magnitude from the adiabatic regime ($C \ll 1$) to the hydrodynamic regime ($C \gg 1$).

In this work it was shown that zonal flows appear equally distributed as positive and negative zonal potential fluctuations on a whole flux surface with narrow radial extent, where the main spectral power is located below 8 kHz. A contribution at higher frequencies is small but could indicate the existence of a geodesic acoustic mode.

The Reynolds stress is not homogeneously distributed but concentrated in regions with negative normal magnetic curvature κ_n and positive geodesic curvature κ_g. This is similar to the distribution of the turbulent transport, which is plausible by reason of similar conceptual form of both quantities. Also, integrated magnetic shear as well as local magnetic shear have an influence on the tilt of the turbulent structures and, therefore, on the Reynolds stress. It is shown that the Reynolds stress drive is large enough to quantitatively explain the acceleration of the flow. The three-wave interaction is strong during the zonal flow occurrence, and the direct energy transfer from small scale structures into the zonal flow supports the picture of a nonlocal driving mechanism in k-space.

The efficiency of the driving mechanism is determined by the density-potential coupling and, therefore, by the collisionality. By analysing Reynolds stress and pseudo-Reynolds stress, it is found that, for decreasing collisionality, the coupling between density and potential increases. As a result, the nonlinear energy transfer into the zonal flow, as well as the relative spectral power of the zonal flow, strongly increases. The zonal flow contribution to the total turbulent spectral power reaches values of up to 29 %. This is a direct test of a fundamental mechanism in plasma turbulence and also represents a verification of the importance of collisionality for large-scale structure formation in magnetically confined toroidal plasmas.

Contents

Chapter 1

Introduction

In order to achieve large scale energy production from fusion reactions, plasmas with a mix of deuterium and tritium are heated to temperatures of 100 million degrees. These temperatures are needed for the particle to overcome the Coulomb barrier. Such fusion plasmas are confined by strong magnetic fields, build in toroidal geometry in order to avoid losses parallel to the magnetic field lines. Important for a fusion reactor is the confinement of the energy measured with the confinement time τ_E, which, together with plasma density n and temperature T, forms the triple product. For ignition, where a self-sustained burning fusion plasma is reached, this quantity has to fulfil the Lawson criterion $nT\tau_E > 4 \cdot 10^{21}\,\mathrm{m}^{-3}\,\mathrm{keV\,s}$ [1]. However, achieving ignition has been proven to be a quite challenging task. Increased heating power leads to higher temperatures but it also entails turbulent fluctuations and turbulent transport. Turbulence reduces the confinement time and severely limits the performance of a future reactor.[1]

The search for improved confinement regimes, which will bring the plasma closer to ignition, has long been subject to fusion research [3, 4]. Enhanced performance modes can be achieved due to specific heating or fuelling scenarios and careful wall preparation [5]. With peaked density profiles the density gradient decay length is reduced below a critical value which can stabilise ion temperate gradient instabilities. In 1982 a new type of improved confinement was discovered at the ASDEX tokamak where the transition to a high confinement regime (H-mode) appeared spontaneously [6–8]. Due to a transport barrier in the edge of the confined plasma [9], the turbulent transport was strongly reduced and the energy confinement time doubled. The transport reduction can be explained by a flow shear layer which hinders turbulent outward transport due to the decorrelation of turbulent struc-

---

[1]The confinement time scales negative with heating power ($P^{-0.5}$) [2] which results in a low confinement regime (L-mode) and leaves the triple product mostly unchanged.

tures (BDT-criterion) [10–13].[2] But the mechanism behind the occurrence of the bifurcation in the confinement time is still not fully understood. Turbulence self-generated zonal flows might play an important role in this H-mode transition [14–18]. These transient shear flows can partially suppress turbulent transport which would in turn increase the ion pressure gradient and thus the background shear flow connected with it. Such behaviour, with a limit cycle oscillation in the intermediate phase between the low and high confinement modes, was indeed confirmed by many experiments [17, 19–25]. However, the physics behind the LH-transition remains a controversial issue as it was found in [26, 27] that the measured turbulent drive was too small to accelerate a zonal flow. And a more resent study suggests that the turbulence zonal flow interaction might not substantially contribute to the LH-transition [28]. Owing to these contrary positions a deeper understanding of the physics related to the drive of zonal flows is highly desirable.

Zonal flows are a phenomenon known before from fluid turbulence [29–31]. The band like structures on Jupiter are probably the most prominent example, but zonal flows also appear in the earth's atmosphere (jet streams) and oceans. On Venus such jets can exhibit velocities faster than the rotation of the planet (super-rotation), and the zonal flows in the interior of the sun are linked to the solar dynamo. Their existence in various physical systems shows that they are a rather universal phenomenon of 2D turbulence.

In toroidal fusion experiments the plasma is confined with an axial (toroidal) magnetic field where the turbulent fluctuations extent far along the field lines and the plasma turbulence is thus quasi two-dimensional. With the additional poloidal magnetic field component, which is needed for stable confinement, the field lines and the elongated turbulent structures are twisted around the torus. The present studies are carried out on a stellarator device where the magnetic field is entirely generated by external field coils and the plasma has a three-dimensional shape. It has been shown that the turbulence in this device resembles that expected in tokamaks, which, in contrast to stellarators, are axisymmetric but need a strong externally induced plasma current to generate part of the field. An illustration of these two confinement concepts is shown in figure 1.1.

Zonal flows exhibit unique properties compared to other turbulent modes. With a homogeneous potential structure along the flux surfaces (called zonal potential) and a finite radial extent zonal flows are intrinsically connected to

---

[2]Turbulent transport is effectively reduced when the shearing rate $\omega_{E \times B}$ is larger than the maximal linear growth rate γ_{max}: $|\omega_{E \times B}| > \gamma_{\mathrm{max}}$.

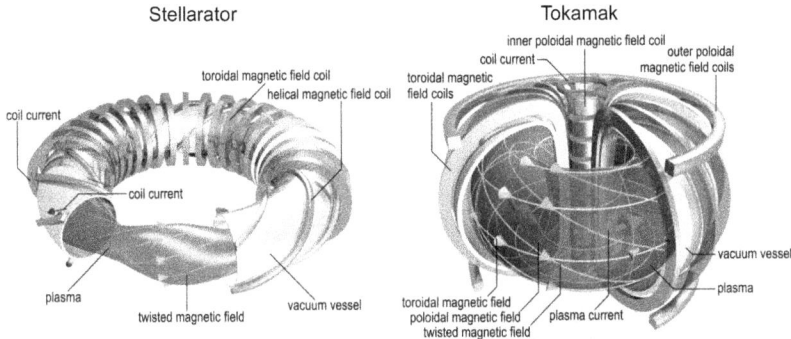

Figure 1.1: Illustration of the two major toroidal magnetic confinement concepts. The combination of toroidal and poloidal magnetic field results in twisted field lines which span the flux surfaces, enclosing volumes of constant magnetic flux. Closed flux surfaces, which do not touch the wall, constitute the confinement region, otherwise the scrape-off layer (SOL). The complex coil geometry of stellarators is reflected in the shape of the plasma. [32]

a zonal shear flow [33]. Because of their symmetry, these mesoscale turbulent structures do not contribute to turbulent cross-field transport and can suppress radial transport by shearing off turbulent eddies. Like in a self-organisation process, the zonal flow is generated by the ambient turbulence itself with a vortex-thinning mechanism [34, 35]. The vortices are tilted and drive the shear flow, which leads to a self-amplification of the zonal flow [36–38]. For tilted vortices the so-called Reynolds stress $\mathcal{R} = \langle \tilde{v}_r \tilde{v}_\theta \rangle$ is non-zero and the radial gradient of this flux surface averaged quantity, as indicated by the brackets, drives the zonal flow. From a theoretical point of view, mostly the physical picture known from fluid mechanics can be transferred to plasma physics. But plasma turbulence, especially in complex geometries, has its own characteristics and phenomena. In contrast to neutral fluids, e.g., the zonal flow drive in plasmas should crucially depend on the cross-coupling of the potential and the density structures. The key parameter in this system is the collisionality C, defined as the normalised electron collision frequency [39]. For the adiabatic case ($C \to 0$) both quantities are closely coupled, while in the hydrodynamic case ($C \to \infty$) density and potential decouple and the zonal flow growth is broken.

The basic concept of large-scale structure formation is clear and elementary processes, like Reynolds stress drive, limit cycle oscillations, and transport suppression, have been demonstrated [40, 41]. However, the experimental verification remains limited as such measurements were mostly restricted to single points in the plasma and do not regard the three-dimensional dynamic of the zonal flow. The complexity of the magnetic field, especially in stellarators, with the consequences for the zonal flow are, up to now, rarely studied in experiment and theory. And a realistic treatment of the geometry in turbulence simulations poses a challenging task which still cannot fully be mastered.

The objective of the present thesis is a detailed analysis of the zonal flows in a toroidally confined plasma with a special focus on the driving mechanism and its collisional dependence. This includes the direct study of the Reynolds stress, and its gradient, together with the connected energy transfer between turbulence and zonal flow. The relevant parameters are measured on the complete poloidal circumference of the confined plasma which allows studying the complex dependency on the magnetic field.

To investigate turbulence, especially the Reynolds stress, multiple measurement points at high time resolution are required which is beyond the limits of the actual diagnostic possibilities in fusion plasmas. Toroidal experiments with low temperature plasmas can fill the gap as their whole confinement region is accessible to probe diagnostics. Probes posses a very high time and a relatively high spatial resolution at the same time. The actual plasma parameters are of course not in the range of those in a fusion plasma, but operation regimes can be chosen such that the normalised parameters relevant for turbulence are comparable to those in the edge region of large fusion experiments. With their flexibility such experiments are predestined for basic research where local magnetic field effects can be very well studied. The experiments for this work have been conducted at the stellarator experiment TJ-K where multi-probe configurations have been exploited to resolve turbulent fluctuations.

This work is organised as follows. The theoretical background of drift waves, the predominant micro instability in the experiment TJ-K, and zonal flows is given in chapters 2 and 3, respectively. This is followed by the description of the techniques of data analysis (Chap. 4), especially the calculation of the energy transfer, and the experimental setup (Chap. 5). All measurements in this work have been performed with newly constructed limiters which result in well-defined boundary conditions. The characterisation of the achieved plasma parameters is presented in chapter 6. This is the foundation for the scaling analysis applied throughout the work. Afterwards the occurrence of the zonal flow is studied in detail where chapter 7 addresses basic properties of the zonal flows. With the conditional averaging technique the temporal evolution can be visualised. In chapter 8 the connection of the Reynolds stress to the magnetic field geometry is studied in detail. Using non-linear analyses techniques, the different energy transfer channels connected with the zonal flow development are studied in chapter 9. Finally, in chapter 10 the results are summarised and discussed with regard to possible consequences of local measurements and the conclusions are presented.

Chapter 2

Plasma turbulence

Turbulence is ubiquitous in nature with a variety of phenomena. This chapter introduces the basic description and characteristics of fluid (Chap. 2.1) and plasma turbulence (Chap. 2.2–2.4). The description is mostly limited to 2D turbulence which is the relevant one for turbulence in fusion plasmas. The fundamental equations introduced here are the basis for the consideration of large structure formation shown in the following chapter.

2.1 Principles of turbulence

In a first part (Sect. 2.1.1) basic formulas, as the Navier-Stokes equation and the vorticity equation, are collected. This is followed by the identification of the conservation laws (Sect. 2.1.2) which entails the turbulent cascades (Sect. 2.1.3) with completely different manifestations in two and three dimensions.

2.1.1 Basic equations

The Navier-Stokes equation [42, 43] is the momentum balance equation for a Newtonian fluid, which, for a complete description, has to be complemented by the continuity equation and an equation for the energy. For an incompressible fluid, with constant mass density and viscosity, it is an extension of the Euler equation by internal friction and describes the evolution of a fluid element in a divergence free velocity field $\mathbf{v}$,

$$D_t \mathbf{v} \equiv \frac{\partial}{\partial t}\mathbf{v} + \mathbf{v} \cdot \nabla \mathbf{v} = -\nabla p + \mu \nabla^2 \mathbf{v} , \tag{2.1}$$

$$\nabla \cdot \mathbf{v} = 0 . \tag{2.2}$$

The mass density ρ_m is thereby included in the pressure p and in the (kinematic) viscosity μ. The differential operator D_t is the hydrodynamic derivative, or material derivative, and describes the rate of change in the co-moving frame of reference. This system of second order nonlinear partial differential equations has to be supplemented by appropriate boundary and initial conditions for the velocity and pressure field, which, at least for two dimensions [44], determine a unique solution.[1] If the nonlinear convective term (second term on the left hand side) can be neglected, e.g. if the viscosity is very high, equation (2.1) reduces to a simple diffusion equation (Stokes equation) and a number of special solutions can be obtained (Stokes or creeping flow [45]). But for the majority of more general flows the nonlinear term is essential to the dynamics of the flow. The relative strength of this convective term in comparison to the viscous term finally determines the state (laminar or turbulent) of the flow. As only a dimensionless control parameter can be of fundamental significance, the viscosity has to be normalised to a typical length L and velocity V of the system, leading to the Reynolds number

$$Re = \frac{L V}{\mu} \,.$$
(2.3)

For low Reynolds numbers momentum diffusion by viscosity dominates and the flow is laminar. With increasing Reynolds number the momentum convection gains importance, which leads to the excitation of a few unstable modes with specific flow pattern like, e.g., a Kármán vortex street of alternating vortices. The number of excited modes gets larger with increased control parameter. Eventually, they get nonlinearly unstable and will finally lead to chaotic behaviour and turbulence.[2]

The parameters used in the definition of the Reynolds number (2.3) also define the possible scales of the turbulence. For large structures the typical geometrical size L defines the integral scale where energy is introduced into the system. On the other hand, the viscosity sets a limit for the size of the small structures, i.e. the Kolmogorov dissipation scale. Due to the Laplace operator in the viscous term, viscous diffusion strongly gains influence for smaller structure sizes where the energy is then dissipated into heat.

A characteristic of turbulent flows is that they are rotational. Therefore, the vorticity $\boldsymbol{\Omega}$, defined as rotation of the velocity field,

---

[1] For three dimensions, the existence and smoothness of a solution is not yet proven and is one of the 'Millennium Problems' announced by the Clay Mathematics Institute.

[2] Different mechanisms for the onset of turbulence are known but the exact route is yet unclear. In the development of drift-wave turbulence the Ruelle-Takens scenario [46] was confirmed [47–49].

$$\boldsymbol{\Omega} = \nabla \times \mathbf{v} \,, \tag{2.4}$$

plays an important role in the study of turbulence and describes the rotation of fluid elements about their centroid. An evolution equation for the vorticity can be deduced from the Navier-Stokes equation by taking the curl of (2.1). With the vector identity for the convective term,

$$(\mathbf{v} \cdot \nabla)\mathbf{v} = \nabla \frac{\mathbf{v}^2}{2} - \mathbf{v} \times (\nabla \times \mathbf{v}) = \nabla \frac{\mathbf{v}^2}{2} - \mathbf{v} \times \boldsymbol{\Omega} \,, \tag{2.5}$$

equation (2.1) leads to

$$\frac{\partial}{\partial t}\boldsymbol{\Omega} = \nabla \times (\mathbf{v} \times \boldsymbol{\Omega}) + \mu \nabla \times \Delta\mathbf{v} \,. \tag{2.6}$$

Because of $\nabla \times (\nabla(\mathbf{v}^2/2 + p)) = 0$, the pressure has been eliminated from the equation. The first term on the right hand side can be simplified with $\nabla \cdot (\nabla \times \mathbf{v}) = 0$ and the incompressibility condition (2.2) to

$$\begin{aligned} \nabla \times (\mathbf{v} \times \boldsymbol{\Omega}) &= \mathbf{v}(\nabla \cdot \boldsymbol{\Omega}) - \boldsymbol{\Omega}(\nabla \cdot \mathbf{v}) + (\boldsymbol{\Omega} \cdot \nabla)\mathbf{v} - (\mathbf{v} \cdot \nabla)\boldsymbol{\Omega} \\ &= (\boldsymbol{\Omega} \cdot \nabla)\mathbf{v} - (\mathbf{v} \cdot \nabla)\boldsymbol{\Omega} \,. \end{aligned} \tag{2.7}$$

Also the third term of equation (2.6) (with the viscosity μ) can be reformulated, using incompressibility, $\nabla \cdot \mathbf{v} = 0$, to

$$\begin{aligned} \nabla \times \Delta\mathbf{v} &= \nabla \times (\nabla(\nabla \cdot \mathbf{v})) - \nabla \times (\nabla \times (\nabla \times \mathbf{v})) \\ &= -\nabla \times (\nabla \times \boldsymbol{\Omega}) = \Delta\boldsymbol{\Omega} \,. \end{aligned} \tag{2.8}$$

With both rearranged terms (Eqs. (2.7) and (2.8)), equation (2.6) results in the vorticity equation in three dimensions

$$\frac{\partial}{\partial t}\boldsymbol{\Omega} + \mathbf{v} \cdot \nabla\boldsymbol{\Omega} = (\boldsymbol{\Omega} \cdot \nabla)\mathbf{v} + \mu\Delta\boldsymbol{\Omega} \,, \tag{2.9}$$

describing the time evolution of the vector $\boldsymbol{\Omega}$. Two terms are originating from the nonlinearity of the Navier-Stokes equation, cf. equation (2.5) and (2.7), which exist also in the absence of viscosity (ideal fluid). The second term of equation (2.9) is the convection of vorticity, and the third term describes the stretching of a vortex line[3], leading to an amplification of the vorticity. This

---

[3] Curves defined as everywhere tangential to the vorticity vector.

vorticity amplification is a consequence of the conservation of circulation Z (Helmholtz's theorem [50]) for ideal fluids,

$$Z = \oint \mathbf{v} \cdot d\mathbf{l} = \int \nabla \times \mathbf{v} \, d\mathbf{S} = \text{const} . \tag{2.10}$$

Thereby, the integration path of the line integral follows a closed vortex line moving with the fluid. The second part of the equation, after Stokes' law is applied, motivates the use of the vorticity, originally defined in (2.4). From equation (2.10) it is now clear that if the cross section of a vortex $\mathbf{S}$ is reduced through convection of the flow, the vorticity has to increase in order to keep the circulation constant. This mechanism produces intense, fine-scale structures as indeed observed in turbulence [51, 52].[4] But also in the vorticity equation, the viscous diffusion term (last term in Eq. (2.9)) is present, which counteracts the vorticity amplification and sets a limit to the structure size. For sufficiently small scales the viscosity becomes important, leading to a diffusion which smoothes out the vorticity field and stops the amplification.

For a two-dimensional flow, i.e. $\mathbf{v} = (v_x, v_y, 0)$, the vorticity has only a component perpendicular to the plane $\mathbf{\Omega} = \nabla \times \mathbf{v} = \Omega \mathbf{e}_z$. Since the derivative of the flow velocity parallel to the vorticity vector is always zero, the first term on the right hand side of equation (2.9) vanishes, and the vorticity equation reduces to

$$\frac{\partial}{\partial t} \mathbf{\Omega} + \mathbf{v} \cdot \nabla \mathbf{\Omega} = \mu \Delta \mathbf{\Omega} . \tag{2.11}$$

In two dimensions, the vorticity equation has reduced to a simple advection-diffusion equation where the vorticity does not act back on the turbulent flow. The missing vorticity stretching is the main difference between two- and three-dimensional turbulence and has, as will be shown later (Sect. 2.1.3), far reaching consequences for the turbulent system.[5]

2.1.2 Conservation laws

Dynamical systems described by the Navier-Stokes equation exhibit deterministic chaos which, in some sense, can be referred to as being sensitive

---

[4] The funnel of a tornado or the vortex above the outlet of a bathtub, also it is a laminar flow, arises due to the same principle.

[5] Some authors suggest that, because of the missing vortex stretching, flows in two dimensions cannot be seen as turbulent systems, and turbulence is intrinsically three-dimensional.

to initial conditions. In principle, the turbulent flow is reproducible, but, since small differences in the initial and boundary conditions quickly lead to completely different flow patterns, only a probabilistic description of the turbulence seems meaningful [53, 54]. The quantities of interest in the study of turbulence are averages over an ensemble of different realisations of an experiment under nominally the same conditions.[6]

In the statistical sense just described fully developed turbulence possesses global symmetries (homogeneity and isotropy). Through Noether's theorem [56] every symmetry is connected to a conserved quantity.[7] Assuming spatial homogeneous turbulence (periodic boundary conditions), or rather symmetry under space-translation, leads to the conservation of momentum

$$\frac{\mathrm{d}}{\mathrm{d}t} \langle \mathbf{v} \rangle = 0 \; . \tag{2.12}$$

Since the Navier-Stokes equation is dissipative, the mean energy $E = \langle 1/2\,\mathbf{v}^2 \rangle$ is only conserved for the inviscid case (Euler equation) and the energy balance equation reads

$$\frac{\mathrm{d}}{\mathrm{d}t} \left\langle \frac{1}{2}\mathbf{v}^2 \right\rangle = -\mu \left\langle |\mathbf{\Omega}|^2 \right\rangle \; ,$$
$$\frac{\mathrm{d}}{\mathrm{d}t} E = -2\mu W \equiv -\epsilon_\mu \; . \tag{2.13}$$

A new quantity called mean enstrophy $W \equiv \langle 1/2\,|\mathbf{\Omega}|^2 \rangle$ is introduced, which captures the energy in the rotation of the flow field. Also the energy dissipation rate ϵ_μ is defined in equation (2.13). It describes the rate at which the system dissipates turbulent kinetic energy into heat at small scales and is important since it is the remaining fluid characteristic determining the turbulent scaling laws (see Sect. 2.1.3).

For the sake of completeness, also the mean helicity[8] $H \equiv \langle 1/2\,\mathbf{v}\cdot\mathbf{\Omega}\rangle$ is a conserved quantity if the viscosity is set to zero [59],

---

[6] With Birkhoff's theorem [55] the ensemble average is connected to an average over time if the system is ergodic. This implies that the time average is the same for (almost) all initial conditions of the system meaning that the system 'forgets' its initial state for sufficiently long times.

[7] In this context global conservation laws are of interest as compared to the more local conservation of the circulation (Eq. 2.10).

[8] The helicity can be interpreted as a measure of the knottedness of vortex lines or the twisting of vortices and is used in the investigation of solar dynamics and the plasma dynamo [57, 58].

$$\frac{\mathrm{d}}{\mathrm{d}t}\left\langle \frac{1}{2}\mathbf{v} \cdot \mathbf{\Omega} \right\rangle = -\mu \left\langle \mathbf{\Omega} \nabla \times \mathbf{\Omega} \right\rangle ,$$

$$\frac{\mathrm{d}}{\mathrm{d}t}H = -2\mu H_{\Omega} ,$$

(2.14)

with the mean vortical helicity defined as $H_{\Omega} \equiv \langle 1/2\,\mathbf{\Omega} \cdot \nabla \times \mathbf{\Omega} \rangle$.

From equation (2.11) it is clear that for ideal two-dimensional turbulence, additionally, the vorticity $\mathbf{\Omega}$ is conserved. Therefore, also the enstrophy obeys a balance equation with the palinstrophy $P \equiv \left\langle \frac{1}{2}|\nabla \times \mathbf{\Omega}|^2 \right\rangle$ which is

$$\frac{\mathrm{d}}{\mathrm{d}t}W = -2\mu P .$$

(2.15)

The enstrophy in two dimensions cannot increase with time, as it does for three dimensions due to the amplification of vorticity (see Sect. 2.1.1). Hence, the enstrophy is forced to decrease because of the viscosity μ, similar as the energy in three dimensions. But equation (2.15) also has important implications for the energy. Since the enstrophy W is bound by equation (2.15) and cannot increase above its initial value, the energy dissipation rate ϵ_μ goes to zero as the viscosity vanishes, i.e. $\epsilon_\mu \xrightarrow{\mu \to 0} 0$. This means that the energy in two-dimensional turbulence is not dissipated by viscosity.

2.1.3 Cascades

The insights presented in section 2.1.2 lead to completely different phenomena in the case of three- and two-dimensional turbulence. The picture of turbulence described so far is based on the assumption that the energy is constantly introduced into the system at some large scale (integral scale) either by a stirring force or instabilities. Due to nonlinear interaction between the turbulent structures, smaller structures are generated and the energy is continuously transmitted down to the smallest ones (Kolmogorov scale) which are determined by viscosity. At the Kolmogorov scale, where the energy input through nonlinear interaction and dissipation by viscosity gets equal, the kinetic energy is finally dissipated into heat. This cascade process (Richardson cascade), where the energy cascades through all the possible scales, is fundamental for turbulence and results in the special shape of turbulent spectra. Normally, it is assumed that the range of scale where the energy is introduced to the system (injection range) is clearly separated, speaking in terms of structure sizes, from the range where it is drained from the system

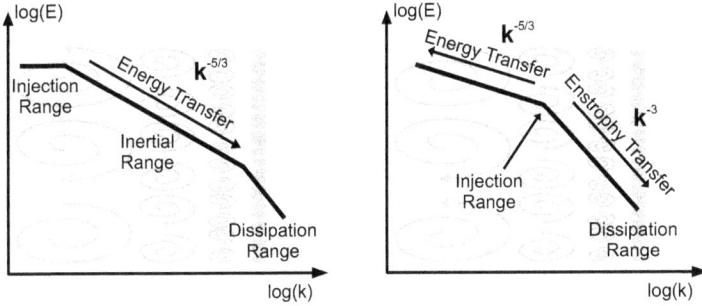

Figure 2.1: Characteristic form of the logarithmic power spectrum of three-dimensional turbulence (left) and two-dimensional turbulence (right). Instead of a single (direct) cascade in three dimensions, the two-dimensional case shows a dual cascade where the energy is inversely transferred to larger scales and the enstrophy is transferred to smaller scales. [60]

(dissipation range). Structures in the injection range and dissipation range then usually differ by orders of magnitude in size. Since we assume homogeneous turbulence, a spectral approach for turbulent flows seems natural where the velocity field is decomposed into its Fourier components $\mathbf{v}_k$,

$$\mathbf{v}(\mathbf{r}) = \sum_k \mathbf{v}_k \exp(i\mathbf{k} \cdot \mathbf{r}) \ . \tag{2.16}$$

But for the description of turbulence at least two point statistics are needed. This is, e.g., the energy, which is of second order, or the often used spatial velocity structure function of the order p, defined as the averaged velocity increment $\delta\mathbf{v}(\mathbf{l})$,[9]

$$S_p(\mathbf{l}) = \langle |\delta\mathbf{v}(\mathbf{l})|^p \rangle \equiv \langle |\mathbf{v}(\mathbf{r}+\mathbf{l}) - \mathbf{v}(\mathbf{r})|^p \rangle \ . \tag{2.17}$$

It can be shown that the second order structure function is directly connected to the spectral energy $E(k) = 1/2 \langle \mathbf{v}_k^2 \rangle$.

For three-dimensional turbulence the power spectrum has the form shown on the left hand side of figure 2.1. The characteristic of this spectrum is that it scales with a constant exponent in the inertial range, which is the region

---

[9]The structure function is normally separated into longitudinal and transverse structure function parallel or perpendicular to the separation l, respectively.

between the injection range on one side and the dissipation range on the other side. Experimentally, one finds two empirical laws for fully developed turbulence [61, 62]:

i. The second order (longitudinal) structure function $S_2(l)$ scales for high Reynolds number turbulence approximately with a two-third power law of the distance l, i.e. $S_2(l) \propto l^{2/3}$.

ii. In the limit of high Reynolds numbers the energy dissipation rate $\epsilon_\mu \equiv - \mathrm{d}/\mathrm{d}t \, E$ (see Eq. (2.13)) becomes independent of viscosity and tends towards a finite value.[10]

In his K41 theory [65] Kolmogorov could show that for homogeneous isotropic turbulence, under the assumption of self-similarity of turbulence and a finite energy dissipation rate, the second order structure function has to scale with the exponent two-third, which reproduces the experimental findings very well.[11] For the spectral energy then follows the scaling factor $-5/3$ resulting in the important scaling relation $E(k) \propto k^{-5/3}$.

The picture is quite different in the case of two-dimensional turbulence which is shown on the right hand side of figure 2.1. Since the vorticity amplification does not take place in two dimensions (see Sect. 2.1.2), the enstrophy cannot increase above its initial value. With the energy balance equation (2.13), it is clear that the energy dissipation rate ϵ_μ vanishes in the limit of zero viscosity. The energy cannot cascade to small scale structures since the viscosity is no sink for the energy, as in contrast to three-dimensional turbulence. Kraichnan [66, 67] followed that the energy, injected at an intermediate scale, has to cascade to larger scales where vortices of the size of the volume (or the box for periodic boundary conditions) finally dissipate due to friction at the wall. But the scaling of this inverse energy cascade should follow the same scaling law as in three-dimensional turbulence, i.e. $E(k) \propto k^{-5/3}$. In contrast to the energy, the enstrophy can be dissipated by viscosity important at small scales. The exact scaling exponent for the direct enstrophy cascade is not as easily predicted as for the energy cascade, and Kraichnan deduced an energy spectrum of the small scales proportional to k^{-3}.

---

[10] This is called the dissipation anomaly [63, 64].

[11] For the third order longitudinal structure function $S_3(l)$ a non-trivial and universal relation can be derived, which is known as the four-fifth law [63].

2.2 Plasma micro-instabilities

A multitude of micro-instabilities exists in magnetised plasma, which can be categorised into two main classes by their structure parallel to the magnetic field. The ideal interchange instability is homogeneous along the magnetic field line and only the dynamics in the cross section perpendicular to the magnetic field determines the stability. On the other hand, the drift-wave (or better drift-wave like) instability is localised in parallel direction to the magnetic field, and the (dominantly) parallel motion of the electrons strongly influences the perpendicular dynamics.

From there on, different modes can be distinguished depending on their underlying driving mechanism. In a toroidal fusion experiment density n and temperature T decline eventually and the resulting pressure gradient $\nabla p = T\nabla n + n\nabla T$ is the source of free energy which drives the linear instabilities. Temperature gradient driven modes, either ion (ITG) or electron temperature (ETG), are normally stronger in the core of the plasma, whereas drift waves, driven by the density gradient, dominate the plasma edge. Whether the instabilities finally get unstable or not depends, among other things, on the background magnetic field structure.

Because of turbulence, the major parameters important for the plasma are not constant in time but fluctuate around their mean value (denoted with the subscript $_0$),

$$n(\mathbf{r}, t) = n_0(\mathbf{r}) + \tilde{n}(\mathbf{r}, t) , \qquad \phi(\mathbf{r}, t) = \phi_0(\mathbf{r}) + \tilde{\phi}(\mathbf{r}, t) ,$$

$$T(\mathbf{r}, t) = T_0(\mathbf{r}) + \tilde{T}(\mathbf{r}, t) , \qquad B(\mathbf{r}, t) = B_0(\mathbf{r}) + \tilde{B}(\mathbf{r}, t) .$$

If magnetic field fluctuations $\tilde{B}$ are small, the electric field $\tilde{E}$ is solely determined by the potential fluctuations, i.e. $\tilde{E} = -\nabla\tilde{\phi}$, and the turbulence is electrostatic.[12] In magnetised plasmas an electric field results in an $E{\times}B$-drift velocity

$$\mathbf{v}^{E{\times}B} = \frac{\mathbf{E} \times \mathbf{B}}{B^2} . \qquad (2.18)$$

That is why a potential perturbation is connected to a vortex structure in the plane perpendicular to the magnetic field. Despite of a pressure gradient, such a structure would in total not result in a net radial transport as it is advecting density in equal parts to the inside and to the outside. The

---

[12] For low-temperature plasmas, like in the experiment TJ-K, also the temperature fluctuations $\tilde{T}$ are small and can be neglected [68].

turbulent transport is given as the product of density $\tilde{n}$ and radial velocity fluctuation $\tilde{v}_r$,

$$\Gamma = \langle \tilde{n}\tilde{v}_r \rangle \ . \tag{2.19}$$

For minimal transport Γ the phase difference between density and velocity fluctuation has to be $\pi/2$ [69, 70], which means zero phase shift for density and potential fluctuation. This becomes clear when the correlation between density and velocity fluctuation, defining the turbulent transport, is written in spectral terms (see Sect. 4.3.1). With the Wiener-Khinchin theorem the cross-correlation function can be connected to the cross power spectrum [71, 72], consisting of the cross coherence $\gamma_{n,v_r}(f)$ between the fluctuations, the respective auto power spectra $S_n(f)$ and $S_{v_r}(f)$, and the cross-phase spectrum $\alpha_{n,v_r}(f)$:

$$\Gamma = \sum_f \gamma_{n,v_r}(f)\sqrt{S_n(f)S_{v_r}(f)}\cos(\alpha_{n,v_r}(f)) \ . \tag{2.20}$$

A high coherence implies a linear dependence of the two signals, indicating a constant phase relation. As the radial drift velocity includes the gradient of the potential (Eq. (2.18)), the cross-phase between density and potential fluctuation $\alpha_{n,\phi}$ is shifted by $\pi/2$, i.e. $\alpha_{n,v_r} = \alpha_{n,\phi} - \pi/2$. Equation (2.20) can be expressed as

$$\Gamma \propto \sum_f \gamma_{n,\phi}(f)\sqrt{S_n(f)S_\phi(f)}\sin(\alpha_{n,\phi}(f)) \ . \tag{2.21}$$

When density and potential are in phase the turbulent structure will not contribute to turbulent transport.

In the following, the underlying mechanism leading to the interchange instability and the drift-wave instability will be discussed in more detail.

2.2.1 Interchange instability

The interchange instability in plasmas is equivalent to the Rayleigh-Taylor instability[13] and requires a pressure gradient and (for the discussion herein) a magnetic field gradient[14]. For this instability the pressure perturbation is,

---

[13] This is a most common instability in turbulent systems as it appears in liquids under gravity or nebulas from supernova explosions.

[14] The discussion in terms of the magnetic field curvature leads to the same result but here the fluid picture is used.

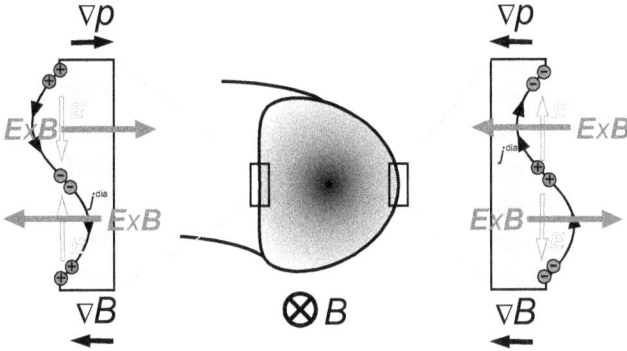

Figure 2.2: Illustration of the formation mechanism of the interchange instability. Depending on the magnetic field structure, here depicted for the inner (left) and outer side (right) of a torus, the interchange instability gets unstable.

in principle, constant parallel to the magnetic field line, i.e. $k_\parallel = 0$, and the dynamic is restricted to the plane perpendicular to the magnetic field.

The formation mechanism of the interchange instability is schematically shown in figure 2.2 for the two cases where the pressure gradient ∇p is either antiparallel or parallel to the magnetic field gradient ∇B. A sinusoidal pressure perturbation perpendicular to the magnetic field is present in both cases, and the diamagnetic current

$$\mathbf{j}^{\mathrm{dia}} = -\frac{\nabla p \times \mathbf{B}}{B^2} , \tag{2.22}$$

runs either to the bottom or the top, respectively. Due to the perturbation, the isobar extends over regions with lower and higher magnetic field as the magnetic field increases to the left. Because the diamagnetic current depends inversely on the magnetic field, the current along the isobar is not constant anymore and charge density develops, $\partial_t \rho = -\nabla \mathbf{j}^{\mathrm{dia}} \neq 0$. For the case with antiparallel orientation (left) this leads to the shown charge accumulation where the resulting $E \times B$-drift (Eq. (2.18)) acts against the pressure perturbation and stabilises it. For this reason such a configuration is called the good curvature region. The situation changes in the opposite case with parallel orientation of pressure and magnetic field gradient (right). Now the resulting $E \times B$-drift is directed in the direction of the perturbation which will be amplified.[15] The phase difference between density and resulting potential fluctuation is $\alpha_{n,\phi} = \pi/2$, which would lead to a high outward transport (cf. Eq. (2.21)). This configuration of pressure and magnetic field gradient is, therefore, called bad curvature region.

In the scrape-off layer (SOL), area outside the confinement region, the field lines end on the wall and the connection length[16] is relatively short, providing the boundary conditions needed for this instability. This leads to the dominance of interchange characteristics in the turbulent fluctuations in the SOL. Plasma blobs, one type of interchange instability, are ejected towards the wall and make major contributions to the turbulent transport [73–78].

In the confined region of the plasma the conditions for ideal interchange modes are not really met. Nevertheless, turbulence in the confinement region can also exhibit interchange like characteristics. This includes ITG [79, 80] and ETG modes [81, 82] and trapped electron modes (TEM) [83, 84], where magnetically trapped particles lead to a potential distribution similar to the one shown in figure 2.2.

2.2.2 Drift-wave instability

The other fundamental instability is the drift-wave instability [85], originating from a pressure perturbation which is localised along a field line ($k_\parallel \neq 0$ and $k_\parallel \ll k_\perp$). Also the dynamics parallel to the magnetic field line plays now an important role. This results in a simultaneous potential perturbation and a propagation perpendicular to the magnetic field and the pressure gradient.

The initial situation is shown on the left hand side in figure 2.3. A localised density perturbation $\tilde{n}$ leads to a current along the magnetic field line, out of the volume of increased density. Due to the much higher mobility of the electrons, a positive potential perturbation $\tilde{\phi}$ arises with zero phase shift compared to the density perturbation (left). The resulting electrical field counteracts this process until ambipolarity is reached again. As the electrical field is connected to a perpendicular drift (Eq. (2.18)), the perturbation is advected and the drift wave begins to propagate downwards (middle). When finite inertia of the ions is included, an additional polarisation drift

$$\mathbf{v}_i^{\text{pol}} = \frac{m_i}{qB^2} \partial_t \tilde{\mathbf{E}}_\perp \, , \tag{2.23}$$

---

[15]This can lead to radially extended structures called streamer with large radial extent.
[16]Path along the magnetic field line.

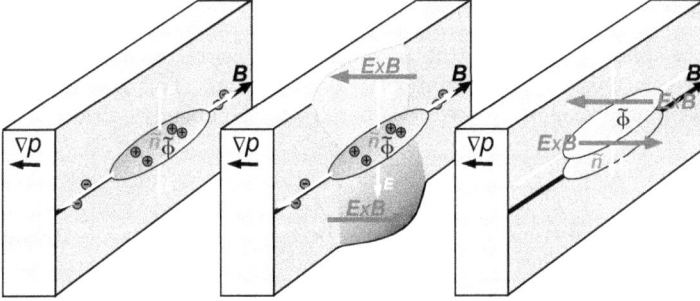

Figure 2.3: Schematic illustration of the formation of a drift-wave instability due to a local pressure perturbation. *Left*: Because of the high mobility, the electrons react fast to the density perturbation $\tilde{n}$ and lead to a positive potential perturbation $\tilde{\phi}$. *Middle*: The arising electrical fields result in $E \times B$-drifts perpendicular to the magnetic field **B** and lead, here, to a propagation of the drift wave downwards. *Right*: The drift waves get unstable when density and potential have a cross-phase unequal zero.

has to be considered, which in turn will counteract the charge separation and slows down the propagation. But the drift wave is then still stable since the phase between density and potential perturbation stays zero. It gets unstable if the parallel electron response is disturbed and the electrons cannot react adiabatically to the density perturbation (right in Fig. 2.3). The electron response can be delayed by numerous factors such as collisions between electrons and ions, induction or Landau damping.

The drift-wave instability is believed to be responsible for a major part of the turbulent transport in the edge of fusion experiments [86, 87] and has found to be the dominating instability in the confined region of the experiment TJ-K [88–91]. A more detailed consideration of this instability is part of the next two sections.

2.3 Drift waves

In this part, the dispersion relation of stable drift waves is deduced (Sects. 2.3.1 and 2.3.2) [60] and mechanisms leading to their destabilisation, and to turbulent transport, are pointed out (Sects. 2.3.3 and 2.3.4).

2.3.1 Dispersion relation of stable drift waves

For the simplest model of drift waves, a density perturbation is assumed to be directly linked to a fluctuation in the potential. This results from the equation of motion of the electrons parallel to the magnetic field,

$$m_e \mathrm{D}_t v_{e\parallel} = e\frac{\partial \phi}{\partial z} - \frac{1}{n_e}\frac{\partial p_e}{\partial z} \ . \tag{2.24}$$

For adiabatic electrons the inertia can be neglected, i.e. $m_e \rightarrow 0$, and the left hand side vanishes. The electric field balances the pressure gradient and, as the temperature is considered constant, $T = T_0$, equation (2.24) reads

$$e\frac{\partial \phi}{\partial z} = \frac{T_e}{n_e}\frac{\partial n_e}{\partial z} \ . \tag{2.25}$$

With integration of equation (2.25) the Boltzmann relation for electrons is recovered,

$$n_e = n_{e0}e^{\frac{e\phi}{T_e}} \ . \tag{2.26}$$

The discussion is restricted to low frequency oscillations where the plasma approximation holds and $n_e = n_i = n$ can be assumed. When only small density fluctuations are considered, $n = n_0 + \tilde{n}$, and higher order terms in the Taylor expansion of the exponential function are dropped, equation (2.26) leads to[17]

$$\frac{\tilde{n}}{n_0} = \frac{e\tilde{\phi}}{T_e} \ . \tag{2.27}$$

Equation (2.27) is the linearised Boltzmann relation where potential fluctuations are directly coupled to density fluctuations.

As described in section 2.2.2, the drift wave evolves from a localised density perturbation. The starting point for the derivation of the dispersion relation is, therefore, the continuity equation[18]

$$\mathrm{D}_t n = \left(\frac{\partial}{\partial t} + \mathbf{v} \cdot \nabla\right) n = -n\nabla \cdot \mathbf{v} \ . \tag{2.28}$$

---

[17]Here, $\phi_0 = 0$ is assumed for the equilibrium plasma.
[18]If the fluid is incompressible, i.e. the density is constant, equation (2.2) follows.

When the ions are considered cold, i.e. $T_i \approx 0$, the velocity $\mathbf{v}$ is reduced to the $E \times B$-drift velocity $\mathbf{v}^{E \times B}$ (Eq. (2.18)).[19] [20] The right hand side of equation (2.28) can be rewritten to

$$\nabla \cdot \mathbf{v}^{E \times B} = \nabla \cdot \left(\frac{\mathbf{E} \times \mathbf{B}}{B^2} \right) = -\mathbf{E} \cdot \left(\frac{\nabla \times \mathbf{B}}{B^2} \right) + \mathbf{B} \cdot (\nabla \times \mathbf{E}) = 0 \ . \quad (2.29)$$

This expression is zero since only a temporal constant ($B(t) = \text{const}$) and homogeneous magnetic field ($\mathbf{B} = B\mathbf{e}_z$) is assumed here. The $E \times B$-drift velocity has only non zero components in the x and y-direction, i.e. $\mathbf{v} = (v_x, v_y, 0)$, and, without loss of generality, the density gradient is set to point into the negative x-direction. As the x-component of the $E \times B$-drift may be written as $v_x^{E \times B} = E_y / B = -(\partial/\partial y)\phi/B$ and the turbulent signals can be separated into mean and fluctuating part, the continuity equation leads to

$$\frac{\partial}{\partial t} \frac{\tilde{n}}{n_0} + \frac{T_e}{eL_n B} \frac{\partial}{\partial y} \frac{e\tilde{\phi}}{T_e} = 0 \ , \quad (2.30)$$

where second order terms have been neglected. Here, the gradient decay length $L_n = -n_0/(\partial n_0/\partial x)$ was introduced. Through the Boltzmann relation derived above (Eq. (2.27)), potential fluctuations can be related to density fluctuations. For a harmonic perturbation in density and potential, i.e. $\tilde{n}, \tilde{\phi} \propto \exp(i(\mathbf{kr} - \omega t))$ with $\mathbf{r} = (x, y, z)$, equation (2.30) leads to

$$\left(\omega - \frac{T_e}{eL_n B} k_y \right) \frac{e\tilde{\phi}}{T_e} = 0 \ . \quad (2.31)$$

This is finally the linear dispersion relation for drift waves:

$$\omega = \omega^{\text{dia}} = \frac{T_e}{eL_n B} k_y \ . \quad (2.32)$$

Phase and group velocity correspond to the diamagnetic drift velocity $\mathbf{v}^{\text{dia}} = -(\nabla p \times \mathbf{B})/(e n B^2)$. With this velocity the drift wave propagates perpendicular to the magnetic field and density gradient into the drift direction of the electrons.

---

[19] The parallel ion dynamic and other drifts are neglected.
[20] The hydrodynamic derivative D_t is then called advective derivative $D_t^{E \times B}$.

2.3.2 Influence of finite ion mass

The electrical fields connected with a drift wave also give rise to a polarisation drift, which counteracts the potential perturbation. Due to the high ion mass this effect is generally small, but for small structure sizes it has to be taken into account.

Since the elongation of the drift wave along the magnetic field line is much larger than its perpendicular extent ($k_\parallel \ll k_\perp$), the parallel charge compensation due to the ion dynamics, i.e. the ion sound wave, is not considered. The polarisation drift of the ions (Eq. (2.23)) is added to the $E \times B$-drift velocity on the right hand side of equation (2.28). With this correction the divergence of the velocity does not vanish anymore and, if higher order terms are neglected, it will result in

$$-n\nabla \cdot \mathbf{v} \approx -n_0 \frac{m_i}{eB} \nabla \cdot \left(\frac{\partial}{\partial t} \tilde{\mathbf{E}}_\perp \right) = n_0 \rho_s^2 \frac{\partial}{\partial t} \left(\nabla_\perp^2 \frac{e\tilde{\phi}}{T_e} \right) . \tag{2.33}$$

$\rho_s = \sqrt{m_i T_e}/eB$ is called the drift scale, which is commonly used in plasma turbulence to obtain dimensionless parameters. Like before, the assumption of a harmonic perturbation is made and an additional term appears in comparison to (2.31),

$$\left(\omega - \frac{T_e}{eL_n B} k_y + \omega(\rho_s k_y)^2 \right) \frac{e\tilde{\phi}}{T_e} = 0 . \tag{2.34}$$

Solving this equation for ω leads to the modified dispersion relation

$$\omega = \frac{\rho_s c_s}{L_n} \frac{k_y}{1 + (\rho_s k_y)^2} = \frac{\omega^{\text{dia}}}{1 + (\rho_s k_y)^2} . \tag{2.35}$$

In contrast to equation (2.32), the dispersion relation deviates from a simple linear relation. Especially, when the structure sizes perpendicular to the magnetic field are small, viz. $k_y \rho_s > 0.3$, the influence of the polarisation drift becomes important and the propagation velocity is reduced.

2.3.3 Unstable drift waves

The polarisation drift changes the propagation velocity of the drift wave structures, but the drift wave is still stable. As discussed in section 2.2.2, only a phase shift between density and potential, i.e. $\alpha_{n,\phi} \neq 0$, makes the drift wave unstable and will lead to an exponential growth of the instability.

This idea is the basis for the so-called $i\delta$-model. The exact mechanism for the perturbed parallel electron dynamics is not important for the time being, and the model just assumes a retarded electron answer, reflected in a modified Boltzmann relation

$$\frac{\tilde{n}}{n_0} \approx \frac{e\tilde{\phi}}{T_e}(1 - i\delta) . \tag{2.36}$$

If this relation is used in the derivation of the dispersion relation (Sect. 2.3.2), equation (2.35) is altered to

$$\omega = \frac{\rho_s c_s}{L_n} \frac{k_y}{1 + (\rho_s k_y)^2 - i\delta} \approx \frac{\omega^{\mathrm{dia}}}{1 + (\rho_s k_y)^2}(1 + i\delta) . \tag{2.37}$$

The dispersion relation has now an imaginary part which is the actual growth rate of the drift wave instability. If the phase shift is positive, $\alpha_{n,\phi} > 0$, meaning a retarded electron response, the drift-wave instability will grow exponentially. Whereas in the case of a negative phase shift $\alpha_{n,\phi} < 0$, the instability is damped and, therefore, stable.

2.3.4 Influence of electron collisions

The case where the parallel electron dynamics is disturbed by collisions will be addressed now. The parallel velocity with which the electrons react to a perturbation is calculated from the equation of motion (see Eq. (2.24)). To include the resistivity due to collisions, an additional term has to be added to the equation:

$$m_e D_t v_{e\parallel} = e\frac{\partial\phi}{\partial z} - \frac{1}{n}\frac{\partial p_e}{\partial z} - m_e \nu v_{e\parallel} . \tag{2.38}$$

The variable ν denotes the electron collision frequency. In the stationary case (electrostatic limit) where the electron dynamic is fast in comparison with the turbulent fluctuations ($m_e \to 0$), it follows from equation (2.38) with a spectral ansatz and neglecting higher order terms:

$$e n_0 i k_\parallel \tilde{\phi} - T_e i k_\parallel \tilde{n} - n_0 m_e \nu v_{e\parallel} = 0 . \tag{2.39}$$

From this equation the parallel electron velocity $v_{e\parallel}$ can be deduced, which is

$$v_{e\parallel} = \frac{i k_\parallel T_e}{m_e \nu}\left(\frac{e\tilde{\phi}}{T_e} - \frac{\tilde{n}}{n_0}\right) . \tag{2.40}$$

The polarisation drift v^{pol} can be neglected for electrons since it is proportional to the mass (cf. Eq. (2.23)), and the divergence of the $E \times B$-drift is zero anyway (Eq. (2.29)). Therefore, only the parallel velocity in equation (2.40) will remain on the right hand side of the continuity equation of the electrons (cf. Eq. (2.28)), which than simplifies to[21]

$$\frac{\partial}{\partial t} \tilde{n} + v_x^{E \times B} \cdot \frac{\partial}{\partial x} n_0 = -n_0 \frac{\partial}{\partial z} v_{e\parallel} . \tag{2.41}$$

For harmonic fluctuations and $\omega^{\text{dia}} = T_e/(eL_n B) \cdot k_y$, equation (2.41) results in

$$\frac{\tilde{n}}{n_0} = \frac{e\tilde{\phi}}{T_e} \left(\frac{\omega^{\text{dia}} + ik_\parallel^2 D_\parallel}{\omega + ik_\parallel^2 D_\parallel} \right) . \tag{2.42}$$

The variable $D_\parallel = T_e/(m_e \nu)$ was introduced in the style of a parallel diffusion coefficient. If it is assumed that $k_\parallel^2 D_\parallel \gg \omega$, equation (2.42) can be further simplified and results in

$$\frac{\tilde{n}}{n_0} = \frac{e\tilde{\phi}}{T_e} \left(1 - i \frac{\omega^{\text{dia}} - \omega}{k_\parallel^2 D_\parallel} \right) . \tag{2.43}$$

This can be compared to the modified Boltzmann relation which was the starting point for the $i\delta$-model (Eq. (2.36)). The phase difference δ is now determined by $D_\parallel$ or rather the collision frequency of the electrons ν. Electron collisions therefore directly lead to a destabilisation of the drift waves.

2.4 Plasma turbulence models

In this section a model for drift-wave turbulence will be developed which describes the temporal evolution of density and potential fluctuations driven by a density gradient [60]. The simplest case with slab magnetic field geometry is presented in section 2.4.1 which is then extended to cover inhomogeneities in the magnetic field (Sect. 2.4.2).

---

[21]The density gradient is again assumed to point into the negative x-direction.

2.4.1 Hasegawa-Wakatani model

To model drift-wave turbulence, the dynamics perpendicular and parallel to the magnetic field have to be connected. The starting point is the quasi neutrality condition

$$\nabla \cdot \mathbf{j} = \nabla_\perp \cdot \mathbf{j}_\perp + \nabla_\parallel \cdot j_\parallel = 0 , \tag{2.44}$$

which is split up into the perpendicular and parallel part. Since there is no current connected to the $E \times B$-drift (Eq. (2.18)), contributions to the perpendicular current $\mathbf{j}_\perp$ can only come from the diamagnetic current $\mathbf{j}^{\text{dia}} = -\nabla p \times \mathbf{B}/B^2$ and the polarisation current $\mathbf{j}^{\text{pol}} = en(\mathbf{v}_i^{\text{pol}} - \mathbf{v}_e^{\text{pol}})$. But the divergence of the diamagnetic current vanishes because [22]

$$-\nabla \cdot \left(\frac{\nabla p \times \mathbf{B}}{B^2} \right) = \nabla p \cdot \left(\frac{\nabla \times \mathbf{B}}{B^2} \right) - \mathbf{B} \cdot (\nabla \times \nabla p) = 0 . \tag{2.45}$$

Because of the mass dependence, the polarisation drift of the electrons can be neglected, $\mathbf{j}^{\text{pol}} \approx en\mathbf{v}_i^{\text{pol}}$. Equation (2.44), together with equation (2.23), is then

$$\nabla_\perp \cdot \left(\frac{m_i n}{B^2} \mathrm{D}_t^{E \times B} \mathbf{E}_\perp \right) + \nabla_\parallel \cdot j_\parallel = 0 . \tag{2.46}$$

This can further be simplified when the electric field is expressed with the potential, $\mathbf{E}_\perp = -\nabla_\perp \phi$, which leads to a vorticity equation

$$\frac{m_i n}{B^2} \mathrm{D}_t^{E \times B} \nabla_\perp^2 \phi = -\nabla_\parallel \cdot j_\parallel . \tag{2.47}$$

For the step from equation (2.46) to (2.47), the spatial derivative $\nabla_\perp$ has to swap places with the advective derivative $\mathrm{D}_t^{E \times B}$. This is not possible without neglecting terms with $\nabla_\perp \cdot \mathbf{v}^{E \times B}$, originating from the product rule. In general they are small compared to $\mathbf{v}^{E \times B} \cdot \nabla_\perp$, but especially for a constant and homogeneous magnetic field the divergence of the $E \times B$-drift velocity is zero (Eq. (2.29)). The connection to the vorticity becomes apparent when the vorticity is written as [23]

$$\mathbf{\Omega} = -\nabla \times \mathbf{v}^{E \times B} = -\nabla \times \left(\frac{\mathbf{E} \times \mathbf{B}}{B^2} \right) = -\nabla_\perp \frac{\mathbf{E}_\perp}{B} \mathbf{e}_z = \nabla_\perp^2 \frac{\phi}{B} \mathbf{e}_z . \tag{2.48}$$

---

[22] The same magnetic field geometry is assumed as in section 2.3.1.

[23] For plasma turbulence the vorticity is defined with a minus sign in contrast to the hydrodynamic definition!

As a next step, the density has to be connected to the parallel electron dynamics. The starting point is the continuity equation (2.28) for the electrons. For small density fluctuations, i.e. $\tilde{n} \ll n_0$, the continuity equation simplifies to

$$\mathrm{D}_t^{E \times B}(\overline{n} + \tilde{n}) = n_0 \nabla \cdot \mathbf{v} \, . \tag{2.49}$$

It is distinguished between the mean density n_0 at a position x_0 and the density profile $\overline{n}$ whose gradients have to be taken into account in the advective derivative. For the contributions on the right hand side, the polarisation drift of the electrons can be neglected and the divergence of the $E \times B$-drift is zero anyway (Eq. (2.29)). The only remaining term is the parallel electron velocity,

$$\mathrm{D}_t^{E \times B}(\overline{n} + \tilde{n}) \approx -n \nabla_\parallel \tilde{v}_{e\parallel} \approx \nabla_\parallel \frac{\tilde{j}_{e\parallel}}{e} \, . \tag{2.50}$$

In this approach, possible contributions of the ions to the parallel current have been disregarded. As in the case of the vorticity equation (2.47), the perpendicular motion of the density due to the $E \times B$-advection is coupled to the parallel electron dynamics.

The parallel electron velocity $v_{e\parallel}$, or rather the parallel current $j_{e\parallel} = e n v_{e\parallel}$, is derived from the equation of motion (2.38) under the assumption of a fast electron response as compared to the turbulent fluctuations (i.e. $m_e \to 0$),

$$\tilde{j}_{e\parallel} = \frac{e}{m_e \nu} \nabla_\parallel \cdot \left(\tilde{p}_e - e n_0 \tilde{\phi} \right) \, . \tag{2.51}$$

Equation (2.51) can be used to substitute the parallel current on the right hand side of the vorticity equation (2.47) and the continuity equation (2.50) resulting in the two basic equations of the simple turbulence model,

$$\left(\frac{\partial}{\partial t} + \mathbf{v}^{E \times B} \cdot \nabla \right) (\overline{n} + \tilde{n}) = \frac{1}{m_e \nu} \nabla_\parallel^2 \left(\tilde{p}_e - e n_0 \tilde{\phi} \right) \, , \tag{2.52}$$

$$\frac{m_i n_0}{B^2} \left(\frac{\partial}{\partial t} + \mathbf{v}^{E \times B} \cdot \nabla \right) \nabla_\perp^2 \tilde{\phi} = \frac{1}{m_e \nu} \nabla_\parallel^2 \left(\tilde{p}_e - e n_0 \tilde{\phi} \right) \, . \tag{2.53}$$

The density is now coupled to the evolution of the vorticity via the parallel electron dynamics.

The two model equations (2.52) and (2.53) will be made dimensionless with the following parameters:

$$\hat{n} = \frac{\tilde{n}}{n_0}, \ \hat{\phi} = \frac{e\tilde{\phi}}{T_e}, \ \hat{p} = \frac{\tilde{p}}{n_0 T_e}, \ \hat{\nabla} = \rho_s \nabla, \ \kappa_n = \frac{\rho_s}{L_n}, \ \hat{t} = t\frac{c_s}{\rho_s} \ . \tag{2.54}$$

When the advective derivative $\mathrm{D}_t^{E \times B}$ is multiplied with ρ_s/c_s it will result in

$$
\begin{aligned}
\frac{\rho_s}{c_s} \mathrm{D}_t^{E \times B} &= \hat{\partial}_t + \left(\frac{\rho_s}{c_s} \frac{T_e}{B \rho_s^2} \right) (\mathbf{e}_z \times \hat{\nabla}_\perp \hat{\phi}) \cdot \hat{\nabla}_\perp \\
&= \hat{\partial}_t + (\mathbf{e}_z \times \hat{\nabla}_\perp \hat{\phi}) \cdot \hat{\nabla}_\perp = \hat{\mathrm{D}}_t^{E \times B} \ .
\end{aligned}
\tag{2.55}
$$

The normalised advective derivative $\hat{\mathrm{D}}_t^{E \times B}$ can then be expressed with Poisson brackets $\{\cdot, \cdot\} = \hat{\partial}_x \hat{\partial}_y - \hat{\partial}_y \hat{\partial}_x$,

$$\hat{\mathrm{D}}_t^{E \times B} = \hat{\partial}_t + \left\{ \hat{\phi}, \cdot \right\} \ . \tag{2.56}$$

Since the density profile is assumed along the x-direction, the left hand side of the continuity equation (2.52), where now the normalised advective derivative $\hat{\mathrm{D}}_t^{E \times B}$ is used, can be rewritten as

$$\hat{\mathrm{D}}_t^{E \times B} (\overline{n} + \tilde{n}) = n_0 \left(\hat{\partial}_t \hat{n} + \left\{ \hat{\phi}, \hat{n} \right\} + \kappa_n \hat{\partial}_y \hat{\phi} \right) \ . \tag{2.57}$$

Hence, the continuity equation (2.52) (first model equation) is

$$\hat{\partial}_t \hat{n} + \left\{ \hat{\phi}, \hat{n} \right\} + \kappa_n \hat{\partial}_y \hat{\phi} = \frac{eB}{m_e \nu} \hat{\nabla}_\parallel^2 \left(\hat{n} - \hat{\phi} \right) \ , \tag{2.58}$$

when multiplied by $\rho_s/(c_s n_0)$. The analogue procedure in the case of the vorticity equation (2.53) (multiplication with $(\rho_s^3 e B^2)/(c_s n_0 m_i T_e)$) will normalise the second model equation,

$$\hat{\partial}_t \hat{\Omega} + \left\{ \hat{\phi}, \hat{\Omega} \right\} = \frac{eB}{m_e \nu} \hat{\nabla}_\parallel^2 \left(\hat{n} - \hat{\phi} \right) \ . \tag{2.59}$$

Here, the dimensionless collision frequency $\hat{\nu}$, normalised to the electron gyrofrequency $\omega_{ce} = eB/m_e$, can be inserted,

$$\hat{\nu} = \frac{\nu}{\omega_{ce}} \ . \tag{2.60}$$

Equations (2.58) and (2.59) together are then the Hasegawa-Wakatani equations in three dimensions [92, 93],

$$\partial_t \hat{n} + \left\{ \hat{\phi}, \hat{n} \right\} + \kappa_n \partial_y \hat{\phi} = \frac{1}{\hat{\nu}} \hat{\nabla}_\parallel^2 \left(\hat{n} - \hat{\phi} \right) \,, \tag{2.61}$$

$$\partial_t \hat{\Omega} + \left\{ \hat{\phi}, \hat{\Omega} \right\} = \frac{1}{\hat{\nu}} \hat{\nabla}_\parallel^2 \left(\hat{n} - \hat{\phi} \right) \,. \tag{2.62}$$

The nonlinearity is included in the Poisson brackets where $\{\hat{\phi}, \hat{n}\}$ incorporates the density advection by the $E \times B$-drift and $\{\hat{\phi}, \hat{\Omega}\}$ originates from the polarisation drift. The density gradient, represented by κ_n, is the source of free energy for the turbulence. Via the right hand side, equations (2.61) and (2.62) are coupled, which represents the intrinsic connection of density and vorticity, or rather potential, in plasma turbulence. But this coupling changes with the collision frequency $\hat{\nu}$, which will be shown to be important for the growth of zonal flows (see Sect. 3.2).

The right hand side of the Hasegawa-Wakatani equations still includes the parallel derivative, which makes them three-dimensional. With the density decay length in parallel direction $L_\parallel$, or alternatively the parallel wavenumber $k_\parallel$, the parallel gradient can be written as

$$\frac{1}{\hat{\nu}} \hat{\nabla}_\parallel^2 \approx \frac{\rho_s^2}{L_\parallel^2 \hat{\nu}} = \frac{(k_\parallel \rho_s)^2}{\hat{\nu}} = \frac{\hat{k}_\parallel^2}{\hat{\nu}} = C^* = \frac{1}{C} \,. \tag{2.63}$$

The dimensionless parameter C^* is the adiabaticity[24], which is the inverse of the collisionality C. With the approximation (2.63), the Hasegawa-Wakatani equations (2.61) and (2.62) can be rewritten containing only parameters in the perpendicular cross section of the magnetic field,

$$\partial_t \hat{n} + \left\{ \hat{\phi}, \hat{n} \right\} + \kappa_n \partial_y \hat{\phi} = C^{-1} \left(\hat{n} - \hat{\phi} \right) \,, \tag{2.64}$$

$$\partial_t \hat{\Omega} + \left\{ \hat{\phi}, \hat{\Omega} \right\} = C^{-1} \left(\hat{n} - \hat{\phi} \right) \,. \tag{2.65}$$

These equations are then the two-dimensional Hasegawa-Wakatani equations.

For an adiabatic electron response the adiabaticity C^* is infinite ($C \to 0$) and the density fluctuations then exactly mimic the potential fluctuations, i.e. $\hat{n} = \hat{\phi}$. In this scenario the Poisson brackets $\{\hat{\phi}, \hat{n}\}$ vanish, and the

---

[24]The nomenclature in the literature is not consistent. Also the adiabaticity is often denoted by C!

right hand side of the Hasegawa-Wakatani equations is zero. When equations (2.64) and (2.65) are subtracted, this will result in the Hasegawa-Mima equation [94]

$$\hat{\partial}_t(\hat{\phi} - \hat{\Omega}) + \kappa_n \hat{\partial}_y \hat{\phi} = \left\{\hat{\phi}, \hat{\Omega}\right\} , \qquad (2.66)$$

or, with the vorticity $\hat{\Omega}$ expressed by the potential (cf. Eq. (2.48)),

$$\hat{\partial}_t(1 - \hat{\nabla}_\perp^2)\hat{\phi} + \kappa_n \hat{\partial}_y \hat{\phi} = \left\{\hat{\phi}, \hat{\nabla}_\perp^2 \hat{\phi}\right\} . \qquad (2.67)$$

In the adiabatic limit density and potential act as single fluid. The Hasegawa-Mima equation is formally the same as the Charney equation [95], which governs, e.g., the dynamics of Rossby waves in the atmosphere [96].

2.4.2 Curvature effects

In the derivation of the Hasegawa-Wakatani model (Sect. 2.4.1) a slab magnetic field geometry was assumed. With this assumption the divergence of the $E\times B$-drift (Eq. (2.29)) and the diamagnetic drift (Eq. (2.45)) vanish. However, this is not the case when the magnetic field is allowed to be inhomogeneous or curved [97–99]. The divergence of the $E\times B$-drift is then

$$\nabla \cdot \mathbf{v}^{E\times B} = \nabla \cdot \left(\frac{\mathbf{E} \times \mathbf{B}}{B^2}\right) = \nabla\phi \cdot \left(\nabla \times \frac{\mathbf{B}}{B^2}\right) . \qquad (2.68)$$

In the case of the electron diamagnetic drift this is

$$\nabla \cdot \left(\frac{\nabla p_e \times B}{enB^2}\right) = -\frac{1}{en}\nabla p_e \cdot \left(\nabla \times \frac{\mathbf{B}}{B^2}\right) . \qquad (2.69)$$

Both drifts determine the perpendicular velocity in the continuity equation (2.50). And the diamagnetic drift adds to the perpendicular current, which leads to an additional term in the vorticity equation (2.47). This results in the modified model equations[25]

$$\mathrm{D}_t^{E\times B}(\overline{n} + \tilde{n}) = \frac{1}{m_e\nu}\nabla_\parallel^2 \left(\tilde{p}_e - en_0\tilde{\phi}\right)$$
$$+ \frac{1}{e}\nabla_\perp(en_0\tilde{\phi} - \tilde{p}_e) \cdot \left(\nabla \times \frac{\mathbf{B}}{B^2}\right) , \qquad (2.70)$$

---

[25] The background density $\overline{n}$ in the diamagnetic drift can be dropped since in a toroidal system the diamagnetic current is in equilibrium with the parallel Pfirsch-Schlüter current.

$$\frac{m_i n_0}{B^2} \mathrm{D}_t^{E \times B} \nabla_\perp^2 \tilde{\phi} = \frac{1}{m_e \nu} \nabla_\parallel^2 \left(\tilde{p}_e - e n_0 \tilde{\phi} \right) + \nabla_\perp \tilde{p}_e \cdot \left(\nabla \times \frac{\mathbf{B}}{B^2} \right) . \quad (2.71)$$

In contrast to the initial model equations (2.52) and (2.53), an additional coupling term appears on the right hand side. In the same manner as in section 2.4.1, the equations are made dimensionless using (2.54). As the magnetic field is still assumed to point in z-direction, $\mathbf{B} = B \mathbf{e}_z$, the normalised geometrical coupling term can be rewritten as

$$\hat{\nabla}(\cdot) \left(\hat{\nabla} \times \frac{\mathbf{B}}{B^2} \right) = \hat{\partial}_x(\cdot) \hat{\partial}_y \ln B - \hat{\partial}_y(\cdot) \hat{\partial}_x \ln B = - \{\ln B, \cdot\} . \quad (2.72)$$

Relation (2.72) together with the modified continuity (2.70) and vorticity equation (2.71) results in the extended (curved) Hasegawa-Wakatani equations

$$\hat{\partial}_t \hat{n} + \left\{ \hat{\phi}, \hat{n} \right\} + \kappa_n \hat{\partial}_y \hat{\phi} = C^{-1} \left(\hat{n} - \hat{\phi} \right) - 2 \left\{ \ln B, \hat{\phi} - \hat{n} \right\} , \quad (2.73)$$

$$\hat{\partial}_t \hat{\Omega} + \left\{ \hat{\phi}, \hat{\Omega} \right\} = C^{-1} \left(\hat{n} - \hat{\phi} \right) - 2 \left\{ \ln B, -\hat{n} \right\} . \quad (2.74)$$

The geometrical coupling term is connected to the magnetic field line curvature $\boldsymbol{\kappa} = (\kappa_n, \kappa_g)$, which has normal and geodesic curvature as components (see Chap. 5.1.2). From the magnetohydrodynamic (MHD) equilibrium condition, $\mathbf{j} \times \mathbf{B} = \nabla p$, follows in a toroidal geometry that

$$\boldsymbol{\kappa} = \frac{\mu_0}{B^2} \nabla_\perp \left(p + \frac{B^2}{2\mu_0} \right) \approx \nabla_\perp \ln B , \quad (2.75)$$

when a small plasma beta $\beta = p/(B^2/2\mu_0) \ll 1$ is assumed [100]. With equation (2.75) the Poisson brackets (2.72) may be written as

$$\{\ln B, \cdot\} = \kappa_n \hat{\partial}_y(\cdot) - \kappa_g \hat{\partial}_x(\cdot) . \quad (2.76)$$

Depending on their sign the curvature components increase the coupling between density and potential.

Chapter 3

Zonal flows

Apart from small scale turbulence, mesoscopic zonal flows are a natural ingredient of turbulence in two dimensions. Like in a self-organisation process, the zonal flows are generated by the ambient turbulence itself and can, therefore, be seen as a secondary instability. The structure of zonal flows in toroidally confined plasmas, and the closely related geodesic acoustic mode, is discussed in section 3.1. Section 3.2 deals with the Reynolds stress drive and the self-amplification process. As drift waves and zonal flows are in a predator-prey like relationship, the dynamics can be described by a Lotka-Volterra model whose different solutions are part of section 3.3. Two important damping mechanisms of zonal flows in toroidal geometry are elucidated in section 3.4.

3.1 Zonal potential modes in plasmas

Zonal flows belong to the group of turbulent modes which are characterised by a zonal potential perturbation [33]. Because of their distinguished spatial structure with a homogeneous potential on a whole flux surface, the direct density response vanishes for such modes. Although the potential might be homogeneous, the connected flow pattern, at least in a toroidal configuration, is not. This leads to a compression of the plasma and, therefore, to a coupling of the potential to the density. In toroidal magnetic field configurations (see Fig. 1.1) such pressure accumulations are balanced along the magnetic field lines as they are twisted around the torus and connect the regions of different pressure.[1] The time it takes for density perturbation to equilibrate in parallel direction is roughly given by the ion sound velocity $c_{\mathrm{s}} = \sqrt{(T_e + T_i)/m_i}$ divided by the connection length $q_{\mathrm{s}} R_0$ (safety

---

[1] This is the same principle as in the case of the Pfirsch-Schlüter currents, where the charge accumulation due to asymmetric diamagnetic currents (Eq. (2.22)) is balanced along the field line.

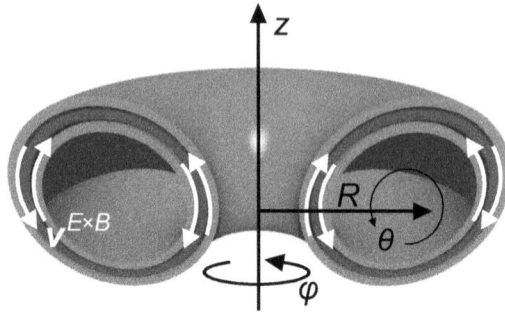

Figure 3.1: Schematic illustration of a zonal flow in toroidal geometry. The negative potential perturbation (blue) is connected to a shear flow, indicated by the arrows [101].

factor q_s [2], major radius R_0). Two regimes can be distinguished. For a slow zonal potential variation, i.e. $\omega_{\text{zonal}} \ll c_s/(q_s R_0)$, the connected flow is incompressible and the toroidicity just leads to a toroidal return flow. This mode is the zonal flow and will be discussed in section 3.1.1 in more detail. If the variation in the potential is comparably fast, i.e. $\omega_{\text{zonal}} \sim c_s/(q_s R_0)$, the result is an oscillation between the zonal potential and the pressure perturbation, which is called the geodesic acoustic mode. The mechanism responsible for the development of this mode will be discussed in section 3.1.2.

3.1.1 Zonal flows

A zonal flow in a magnetically confined toroidal plasma, sketched in figure 3.1, is an extreme example of a convective cell where, in general, the potential structure has zero wavenumber in toroidal direction φ. The convective flow $\mathbf{v}^{E \times B}$ (Eq. (2.18)) connected with the homogeneous potential perturbation is always tangential to the magnetic flux surface. At the same time, the flow has a narrow radial extent with relatively small radial wavenumber.

---

[2] This is the inverse of the rotational transform ι, which is a measure for the twist of the field lines around the torus as it gives the difference in poloidal angle $\Delta\theta$ of a field line after one toroidal turn, i.e. $q_s^{-1} = \iota = \Delta\theta/2\pi$.

Together, the essential zonal flow features are:[3]

$$k_\theta = k_\varphi = 0, \qquad k_r \neq 0 . \tag{3.1}$$

These characteristics imply that the potential perturbation of a zonal flow is always connected to a shear flow, mainly directed in the poloidal direction θ. The toroidicity, which entails a magnetic field strength dependence in the radial direction R, induces the inhomogeneities in the poloidal flow. For a simple torus with $B(R) \propto 1/R$, the flow $(v^{E \times B} \propto E_r/B)$ is faster on the outboard side than on the inboard side, which leads to a compression on the top or bottom. The parallel return flows, which make the total flow divergence free, have the form [102]

$$v_\varphi = \frac{E_r}{B} 2q_s \cos(\theta) , \tag{3.2}$$

depending on the safety factor q_s. For stellarators with their magnetic field ripples, the flow is more complex and does crucially depend on the configuration of the particular experiment.

Due to the vanishing parallel wavenumber, the Boltzmann relation (Eq. (2.27)) does not hold for the zonal potential and, in principle, no density perturbation is connected with the zonal flow. Since there is no parallel acceleration and the linear polarisation drift (cf. Eq. (2.23)) vanishes, the dispersion relation of the zonal flow is especially simple, i.e.

$$\omega = 0 . \tag{3.3}$$

The special structure of the zonal flow has further consequences. Since the potential is homogeneous on the flux surface, Landau damping is especially small. In addition, the zonal flow does not contribute to radial turbulent transport since it has no poloidal electric field components (see Chap. 2.2). But even more important, as it is a poloidal shear flow, the zonal flow can reduce turbulent transport by shearing off turbulent eddies (see Sect. 3.3).

3.1.2 Geodesic acoustic mode

The other branch of the solution for the zonal potential $(m = n = 0)$ is at higher frequencies as compared to the zonal flow, called geodesic acoustic mode (GAM). The potential structure and the connected flow pattern are essentially the same as in the case of the zonal flow but are now accompanied

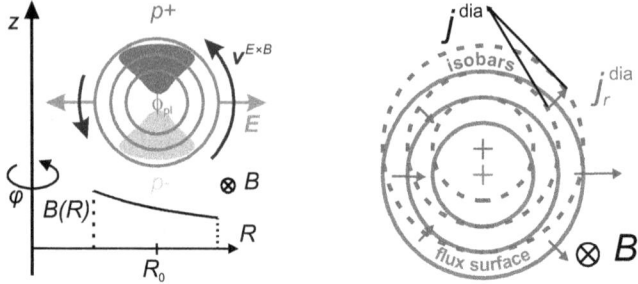

Figure 3.2: Illustration of the mechanism leading to the GAM oscillation. *Left:* Cross section in a simple toroidal geometry. The magnetic field strength $B(R)$ is inversely proportional to the radius. This leads to $E \times B$-drifts ($v^{E \times B} \propto E_r/B$) with different magnitude on the inner and outer side of the torus and a compression at the top or bottom, respectively. *Right:* Due to the displacement of the pressure profile, the diamagnetic current has a radial component j_r^{dia}, which compensates for the asymmetry in the $E \times B$-drift velocity. [103]

by a pressure perturbation. In figure 3.2 the driving mechanism of the GAM is shown schematically. The variation of the magnetic field leads to the asymmetries in the $E \times B$-drift velocity (left figure) and the perpendicular flow is, therefore, not divergence free,

$$
\begin{aligned}
\nabla_\perp \cdot \mathbf{v}^{E \times B} &= -2\,\mathbf{v}^{E \times B} \cdot \nabla_\perp \ln B \\
&= -2\,\mathbf{v}^{E \times B} \cdot \boldsymbol{\kappa} \\
&\propto -\frac{\kappa_{\mathrm{g}}}{B} \,.
\end{aligned}
\tag{3.4}
$$

For a quasi-static variation, like the zonal flow, the resulting pressure perturbation is directly compensated in parallel direction along the field line, i.e. $\nabla_\perp \cdot \mathbf{v}^{E \times B} = -\nabla_\parallel \cdot v$. Otherwise a $m = 1$ density perturbation will persist which will shift the pressure profile (isobars in the right figure) in comparison to the flux surfaces. This leads to diamagnetic currents (Eq. (2.22)) with a non vanishing radial component j_r^{dia}.[4] As it is shown in the figure, these currents counteract the potential perturbation, trying to reverse the

---

[3] In contrast, streamer constitute the opposite case with a large radial extent ($k_r \cong 0$) and a poloidal wavenumber unequal zero ($k_\theta \neq 0$).

[4] Due to this currents, a magnetic field fluctuation is connected to the GAM oscillation, which can be observed in fusion experiments [104, 105].

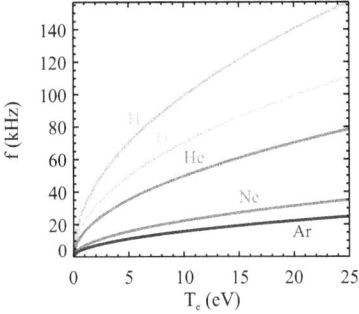

Figure 3.3: Scaling of the GAM frequency with the electron temperature T_e for different gases at TJ-K parameters. The oscillation frequency strongly increases for smaller ion masses. [103]

radial electric field E_r. Because of this restoring force, the perturbation will start to oscillate with the characteristic frequency ω_{GAM}. For the GAM the toroidicity is a prerequisite and, therefore, GAMs cannot develop in simple slab geometry.

In contrast to the zonal flow, the dispersion relation for the GAM is not trivial and leads to the oscillation frequency ω_{GAM}, which, for large aspect ratio $(a \ll R_0)$ and circular plasma, is given by [106]

$$\omega_{\mathrm{GAM}}^2 \simeq \frac{c_{\mathrm{s}}^2}{R_0^2} \left(2 + \frac{1}{q_{\mathrm{s}}^2} \right) . \tag{3.5}$$

The frequency of the GAM is determined by the time it takes for the density perturbation (sound wave) to propagate once around the torus, why geometrical parameters like, e.g., elongation [107] have a strong influence.[5] Characteristic for the GAM is that it scales with the sound velocity c_{s}. In figure 3.3 the simple scaling (3.5) is depicted for typical TJ-K parameters.

The GAM is subject to Landau damping $(\propto \exp(-q_{\mathrm{s}}^2))$ [108–110] and, since the safety factor normally increases with increasing minor radius, typically found in the edge of the confined region. As the zonal flow and the GAM have essentially the same flow structure, both modes can couple. Therefore, the GAM represents a sink for the zonal flow energy, which is a major loss channel for zonal flows in toroidal configurations (see Sect. 3.4.2).

---

[5] Especially for stellarators, the situation is more complicated and an analytic formula is hard to obtain. [108]

3.2 Turbulent Reynolds stress drive

For the description of the zonal flow the poloidal mean or rather the zonal average is of interest. With the Reynolds decomposition, the velocity field and the pressure is split into the mean (denoted by a bar) and the superimposed fluctuating part (denoted by a tilde),

$$\mathbf{v} = \bar{\mathbf{v}} + \tilde{\mathbf{v}} \,, \tag{3.6}$$

$$p = \bar{p} + \tilde{p} \,. \tag{3.7}$$

The mean quantity is generally defined as the ensemble average $\langle \cdot \rangle$, where for now it is not substituted by another, e.g. temporal, average. Mean and fluctuating part fulfil the following relations,

$$\langle \mathbf{v} \rangle = \langle \bar{\mathbf{v}} \rangle = \bar{\mathbf{v}} \,, \qquad \langle \tilde{\mathbf{v}} \rangle = 0 \,. \tag{3.8}$$

Equation (3.8) applies similarly to the pressure p. The different parts of the flow field do not only differ in their statistical distribution, but also vary greatly in scale. While the mean flow changes over relatively long times or large differences, the fluctuations are small scaled, which will become important when derivatives are involved. In here, only incompressible flows are considered,[6]

$$\langle \nabla \cdot \mathbf{v} \rangle = \nabla \cdot \bar{\mathbf{v}} = 0 \,, \qquad \nabla \cdot \tilde{\mathbf{v}} = 0 \,. \tag{3.9}$$

To obtain the evolution of the mean flow, the Reynolds decomposition is used in the Navier-Stokes equation (2.1),

$$\frac{\partial}{\partial t}(\bar{\mathbf{v}} + \tilde{\mathbf{v}}) + ((\bar{\mathbf{v}} + \tilde{\mathbf{v}}) \cdot \nabla)(\bar{\mathbf{v}} + \tilde{\mathbf{v}}) = -\nabla(\bar{p} + \tilde{p}) + \mu \nabla^2(\bar{\mathbf{v}} + \tilde{\mathbf{v}}) \,, \tag{3.10}$$

and the ensemble average is taken. Because of (3.8) and (3.9), the mixed terms in the convective derivative are zero,

$$\bar{\mathbf{v}} \cdot \nabla \langle \tilde{\mathbf{v}} \rangle = \langle \tilde{\mathbf{v}} \rangle \cdot \nabla \bar{\mathbf{v}} = 0 \,. \tag{3.11}$$

This simplification leads to an evolution equation of the mean flow $\bar{\mathbf{v}}$,

$$\frac{\partial}{\partial t}\bar{\mathbf{v}} + (\bar{\mathbf{v}} \cdot \nabla)\bar{\mathbf{v}} + \langle(\tilde{\mathbf{v}} \cdot \nabla)\tilde{\mathbf{v}}\rangle = -\nabla\bar{p} + \mu\nabla^2\bar{\mathbf{v}} \,. \tag{3.12}$$

---

[6]The derivative ∇ commutes with the ensemble average since the incompressible continuity equation (2.2) is linear.

But equation (3.12) includes a part (third term) solely determined by the fluctuations. For the next steps, equation (3.12) is treated in component notation[7], which reads

$$\partial_t \bar{v}_i + (\bar{v}_j \partial_j)\,\bar{v}_i + \langle (\tilde{v}_j \partial_j)\,\tilde{v}_i \rangle = -\partial_i \bar{p} + \mu \partial_j^2 \bar{v}_i \ . \tag{3.13}$$

The third term, which includes the fluctuations, can now be rewritten as

$$\langle (\tilde{v}_j \partial_j)\,\tilde{v}_i \rangle = \partial_j \langle \tilde{v}_i \tilde{v}_j \rangle - \langle \tilde{v}_i (\partial_j \tilde{v}_j) \rangle \ . \tag{3.14}$$

Because of (3.9), the last term in equation (3.14) vanishes. The second-order moment $\langle \tilde{v}_i \tilde{v}_j \rangle$ is then substituted in equation (3.13),

$$\begin{aligned}
\partial_t \bar{v}_i + (\bar{v}_j \partial_j)\,\bar{v}_i &= -\partial_i \bar{p} + \mu \partial_j^2 \bar{v}_i - \partial_j \langle \tilde{v}_i \tilde{v}_j \rangle \\
&= \partial_j \left\{ -\bar{p}\delta_{ij} + \mu \left(\partial_j \bar{v}_i + \partial_i \bar{v}_j\right) - \langle \tilde{v}_i \tilde{v}_j \rangle \right\} \ .
\end{aligned} \tag{3.15}$$

Equation (3.15) is known as the Reynolds equation or Reynolds-averaged Navier-Stokes equations (RANS equations). The terms in the curly brackets on the right hand side are in sequence: the mean pressure tensor, with only elements on the diagonal, the mean viscous stress tensor, originating from the vector Laplacian term, and the Reynolds stress tensor. The Reynolds stress describes a momentum flux from the turbulent fluctuations to the mean flow and introduces a coupling between both parts of the velocity field.[8] Equation (3.15) shows that, in general, the mean flow cannot be described by just considering averaged quantities, but involves a second-order moment of the fluctuating velocity.[9] An equation for the second-order moment can be found, which in turn involves moments of the next higher order. This can be repeated so on and so forth but, since every other equation involves moments of the next order, the set of equations is not closed, which states the so-called closure problem. The set of equations can be closed using a closure approximation [111], however, information of the turbulence will always be lost if not the full statistics of the fluctuations is used.

The Reynolds stress is a tensor of second order with the single point velocity correlations (in the limit of zero separation or time lag) as components.

---

[7]In this notation the abbreviations are $\partial_t \equiv \partial/\partial t$, $\partial_i \equiv \partial/\partial x_i$, and $\partial_{ij} \equiv \partial^2/\partial x_i \partial x_j$.

[8]The product of the velocity fluctuations, originating in the quadratic nonlinearity of the Navier-Stokes equation, becomes a convolution in Fourier space, $\hat{v}_i \otimes \hat{v}_j = \int \hat{v}_i(\mathbf{k}')\hat{v}_j(\mathbf{k} - \mathbf{k}')\,\mathrm{d}^3\mathbf{k}'$. This describes three-wave (triad) interactions as it involves modes at wavenumbers $\mathbf{k}$, $\mathbf{k}'$, and $\mathbf{k} - \mathbf{k}'$.

[9]This is not the case for isotropic turbulence where the second-order moment of the velocity vanishes.

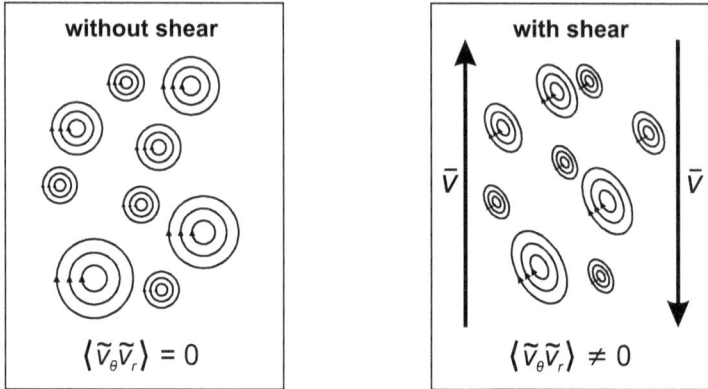

Figure 3.4: The Reynolds stress is zero when the velocity distribution is isotropic (left). With, e.g., a shear flow which tilts the vortices (right), this symmetry is broken and the Reynolds stress takes finite values. [60]

Due to the symmetry of the correlation, the Reynolds stress tensor is symmetric and positive semi definite. As for other stress tensors, the components can be split into the normal stress components $\langle \tilde{v}_i \tilde{v}_i \rangle$ on the diagonal and the off-diagonal shear stresses $\langle \tilde{v}_i \tilde{v}_j \rangle$ acting tangential to the normal directions. When the isotropy of the turbulence is broken, due to inhomogeneities like, e.g., shear flows, the off-diagonal elements are non-zero as it is shown in figure 3.4.

Until now the average $\langle \cdot \rangle$ in (3.8) has not been specified. Here, the zonal flow in torus geometry (r, θ, φ) is of interest, which is characterised as a homogeneous mode on a flux surface (cf. Sect. 3.1.1). Therefore, the average is taken spatially as a poloidal average along a flux surface. For simplicity, the pressure gradient and the viscous term in equation (3.15) are not regarded further. Equation (3.15) is then the averaged poloidal momentum balance equation

$$\frac{\partial}{\partial t} \bar{v}_\theta + (\bar{\mathbf{v}} \cdot \nabla) \bar{v}_\theta = -\partial_j \langle \tilde{v}_\theta \tilde{v}_j \rangle \ . \tag{3.16}$$

Since it is assumed that the zonal flow does not vary in toroidal direction and has no averaged radial velocity component, only the poloidal dependence will persist in the advective term. The right hand side includes the radial gradient of the Reynolds stress tensor component $\langle \tilde{v}_\theta \tilde{v}_r \rangle$. Normally, only

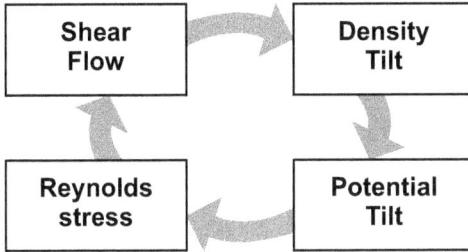

Figure 3.5: Sketch of the self-amplification process connected to the drive of zonal flows. In plasmas turbulence the cross-coupling between density and potential field is involved.

this (perpendicular) Reynolds stress is considered because it is the main component of the Reynolds stress tensor. Nevertheless, there is also the tensor component $\langle \tilde{v}_\theta \tilde{v}_\varphi \rangle$ (parallel Reynolds stress), which defines the averaged poloidal momentum flux due to toroidal velocity fluctuations. If only the main contributions to the poloidal momentum balance are considered and a homogeneous poloidal flow is assumed, i.e. a constant magnetic field, equation (3.16) simplifies to

$$\frac{\partial}{\partial t} \bar{v}_\theta = -\frac{\partial}{\partial r} \langle \tilde{v}_\theta \tilde{v}_r \rangle \ . \tag{3.17}$$

So, the zonal flow is only accelerated when there is a radial gradient in the Reynolds stress $\langle \tilde{v}_\theta \tilde{v}_r \rangle$. The Reynolds stress drive is, therefore, a transfer of turbulent poloidal momentum to the poloidal mean flow.

In plasma turbulence the Reynolds stress drive also involves the cross-coupling between density and potential (Fig. 3.5). The turbulent structures consist of a density and a potential component, which are coupled via the parallel electron dynamics (see Chap. 2.4.1). The shear flow tilts the structures in the density but leaves the potential unaffected at first. Only due to the cross-coupling, the actual vortex flow $\tilde{v} \propto \nabla_\perp \tilde{\phi}$ is tilted, which leads to net Reynolds stress, closing the loop. Therefore, an increased collisionality should hinder zonal flow drive as it reduces the density-potential coupling (cf. Eqs. (2.64) and (2.65)).

The derivation of the Reynolds stress drive equation was done in the electrostatic limit, which is valid for low β-plasmas. In general, also fluctuations in the plasma currents and, therefore, in the magnetic field would have to be considered. The additional term has also the form of a (second order) tensor and is called the Maxwell stress $\langle \tilde{b}_i \tilde{b}_j \rangle$. For drift-Alfvén turbulence the contributions of Reynolds stress and Maxwell stress have opposite sign and can thus reduce the growth rate of the zonal flow [112, 113].

3.3 Drift-wave zonal-flow dynamics

For the description of the zonal flow dynamics in drift-wave turbulence, the evolution of the poloidal shear flow has to be connected to the response of the drift-wave spectrum to the shear (Sect. 3.3.1). Solutions to the resulting predator-prey model are derived in section 3.3.2 and 3.3.3. The presentation follows reference [33].

3.3.1 Model equations

The zonal flow dynamics are governed by the poloidal momentum balance equation (3.17), as the density perturbation can be neglected and the zonal flow is basically a convective cell. Thus, the evolution of the zonal shear, i.e. zonal vorticity $U = v'_{\mathrm{ZF}} = \nabla^2_r \tilde{\phi}_{\mathrm{ZF}}/B$, can be described by

$$\frac{\partial}{\partial t} \nabla^2_r \tilde{\phi}_{\mathrm{ZF}} = -\frac{\partial}{\partial r} \left\langle \tilde{v}_r \nabla^2 \tilde{\phi}_{\mathrm{DW}} \right\rangle - \gamma_{\mathrm{D}} \nabla^2_r \tilde{\phi}_{\mathrm{ZF}} \ , \tag{3.18}$$

where γ_{D} includes generic zonal flow damping.[10] The zonal flow is driven through a radial transport of drift wave vorticity, i.e. $\langle \tilde{v}_r \nabla^2 \tilde{\phi}_{\mathrm{DW}} \rangle$ (cf. Eq. (2.19)), which is, therefore, rather a redistribution than a generation process. The nonlinear term may be expressed in Fourier-space, which gives

$$\frac{\partial}{\partial t} \nabla^2_r \tilde{\phi}_{\mathrm{ZF}} = \frac{1}{B} \frac{\partial^2}{\partial r^2} \int \mathrm{d}^2k \, k_r k_\theta |\tilde{\phi}_{\mathrm{DW},k}|^2 - \gamma_{\mathrm{D}} \nabla^2_r \tilde{\phi}_{\mathrm{ZF}} \ . \tag{3.19}$$

Here, the connection to the drift-wave vortex density or rather energy density $\tilde{N}_{\mathbf{k}} = (1 + k^2 \rho^2_{\mathrm{s}})^2 |\tilde{\phi}_{\mathrm{DW},\mathbf{k}}|^2$ (cf. Eq. (2.35)) can be made,

$$\frac{\partial}{\partial t} \nabla^2_r \tilde{\phi}_{\mathrm{ZF}} = \frac{1}{B} \frac{\partial^2}{\partial r^2} \int \mathrm{d}^2k \, k_r k_\theta \frac{\tilde{N}_{\mathbf{k}}}{(1 + k^2 \rho^2_{\mathrm{s}})^2} - \gamma_{\mathrm{D}} \nabla^2_r \tilde{\phi}_{\mathrm{ZF}} \ . \tag{3.20}$$

Eventually, an equation for the evolution of the zonal flow enstrophy U^2 shall be obtained. When multiplied by U and a coherent response to the

---

[10]This can be a scalar or an integro-differential operator.

effect of the shear flow is assumed, i.e. $\tilde{N}_\mathbf{k} = (\delta N/\delta v'_\mathrm{ZF})v'_\mathrm{ZF}$, equation (3.20) results in

$$\frac{\partial}{\partial t}U^2 = \frac{1}{B^2}\frac{\partial^2}{\partial r^2}\int \mathrm{d}^2k \frac{k_r k_\theta}{(1+k^2\rho_\mathrm{s}^2)^2}\frac{\delta N}{\delta v'_\mathrm{ZF}}U^2 - \gamma_\mathrm{D}U^2 \ . \tag{3.21}$$

The modulation of the drift-wave spectrum $\delta N/\delta v'_\mathrm{ZF}$ leads to an amplification of the shear flow why drift-wave turbulence is unstable to the growth of zonal flows. This can also be shown for a plane drift wave [114], which couples in a four-wave interaction via two sidebands to the $k_\theta = 0$ mode (modulational instability).[11] [12] Both models give the same zonal flow growth rate in the limit when a long-lived primary drift wave is assumed.

For a complete model the response of the drift-wave spectrum to the modulation by the shear flow has to be calculated. This can be derived from the wave kinetic equation[13]

$$\frac{\partial N_\mathbf{k}}{\partial t} - \{\omega, N_\mathbf{k}\} = \gamma_\mathbf{k} N_\mathbf{k} \ , \tag{3.22}$$

expressed with Poisson brackets $\{\cdot,\cdot\} = \partial_\mathbf{x}\partial_\mathbf{k} - \partial_\mathbf{k}\partial_\mathbf{x}$. In quasi-linear approximation and for the case where the lifetime of drift waves and zonal flows is short compared to the characteristic evolution time of the system (time scale of the linear zonal flow instability), it follows[14]

$$\left(\frac{\partial}{\partial t} - \gamma_\mathrm{L} + \gamma_\mathrm{NL}\right)N_\mathbf{k} = \frac{\partial}{\partial k_r}\left(D_k \frac{\partial N_\mathbf{k}}{\partial k_r}\right) \ . \tag{3.23}$$

This describes a diffusion in k-space with the corresponding flux $D_k \frac{\partial N_\mathbf{k}}{\partial k_r}$, induced by the random zonal flow shearing, which, for a stationary state,

---

[11] Still, this involves only three independent waves, interacting in two pairs.

[12] In the process of shearing the other turbulent modes are forced to couple to the zonal flow ($m = 0$) as the resonance manifold of other possible mode interactions shrinks and the respective coupling coefficient is weakened [115]. This is equivalent to the physical picture of the straining-out process [116] where vortices are tilted and coiled up by the shear flow [34]. This manifold shrinking also implies that a background shear flow, which 'pre'-tilts the eddies, leads to a higher zonal flow level.

[13] Known from geometrical optics, it describes the evolution of a distribution of wave packets [117, 118]. Here it can be used since a clear time separation between the low frequency zonal flow and the higher frequency drift waves is assumed, i.e. $\Omega_\mathrm{ZF} \ll \omega_\mathrm{DW}$. Hence, the zonal flow adiabatically modulates the drift-wave spectrum. If this cannot be applied, an envelope formalism has to be used, which does also capture wave diffraction [119].

[14] This corresponds to a system state with stochastically excited modes ('drift wave ray chaos') with random zonal flow shearing. A long lifetime of drift waves and zonal flows would be the opposite case where wave-packet trapping may terminate the zonal flow growth. [33]

has to be in balance with the linear drive $\gamma_{\rm L}$ and the nonlinear damping $\gamma_{\rm NL}$ (self-interaction) of the drift waves. The modulation induced by $v'_{\rm ZF}$ can be deduced from equation (3.23) which, when inserted in equation (3.21), leads to

$$\left(\frac{\partial}{\partial t} + \gamma_{\rm D}\right) U_q^2 = \frac{\partial^2}{\partial r^2} \sum_k D_q \frac{\partial N_{\bf k}}{\partial k_r} U_q^2 . \tag{3.24}$$

In contrast to equation (3.23), the diffusion term (right hand side) is negative consistent with an inverse energy transfer (inverse cascade), but the zonal flow drive is non-local in k-space.[15] The set of equations (3.23) and (3.24) are the model equations for the drift-wave zonal-flow dynamics, which will be further analysed in the following.

3.3.2 Self-consistent states

For a basic analysis of the system dynamics, the number of degrees of freedom can be reduced by using the enstrophy (cf. Chap. 2.1.2) of the drift waves

$$\langle N \rangle = \sum_{k_\theta \neq 0, k_r \neq 0} N_{\bf k} = \sum_{k_\theta \neq 0, k_r \neq 0} (1 + (k\rho_{\rm s})^2)^2 |\phi_{\bf k}|^2 , \tag{3.25}$$

and of the zonal flows

$$\langle U^2 \rangle = \sum_q U_q^2 = \sum_{k_\theta = 0, k_r \neq 0} (k_r \rho_{\rm s})^4 |\phi_{\bf k}|^2 . \tag{3.26}$$

For now the nonlinear damping of the drift waves $\gamma_{\rm NL}$ will be neglected, and the integration of equation (3.23) and (3.24), with use of an approximation for the right hand side, leads to

$$\left(\frac{\partial}{\partial t} - \gamma_{\rm L}\right) \langle N \rangle = -\alpha \langle U^2 \rangle \langle N \rangle , \tag{3.27}$$

$$\left(\frac{\partial}{\partial t} + \gamma_{\rm D}\right) \langle U^2 \rangle = \beta \langle U^2 \rangle \langle N \rangle . \tag{3.28}$$

α and β are constants. This system of differential equations, coupled via the product term on the right hand side, has the form of the Lotka-Volterra

---

[15]This is similar to the result obtained in the eddy-viscosity approximation [120] where the Reynolds stress in equation (3.17) is replaced by a term solely determined by the mean flow velocity, i.e. $-\langle \tilde{v}_\theta \tilde{v}_r \rangle = \mu_{\rm T} \partial_r \bar{v}_\theta$. A negative turbulent viscosity $\mu_{\rm T}$ represents flow drive.

equations [121], which describe the self-regulating dynamics of two populations. In this predator-prey relation one population grows on the cost of the other one.[16] Transferred to the drift-wave zonal flow turbulence, the drift waves $\langle N \rangle$ (primary instability) are the prey where the zonal flow $\langle U^2 \rangle$ (secondary instability) feeds on.

Stationary solutions of this system of equations can be obtained with $\partial \langle N \rangle / \partial t = \partial \langle U^2 \rangle / \partial t = 0$. The first is trivial, whereas it is not stable:

$$\langle N \rangle = 0 , \quad \langle U^2 \rangle = 0 . \tag{3.29}$$

The second stationary solution is the fix point

$$\langle N \rangle = \frac{\gamma_D}{\beta} , \quad \langle U^2 \rangle = \frac{\gamma_L}{\alpha} , \tag{3.30}$$

which is stable and describes a state of the system where the growth and decay of both populations is just in balance.

With a constant of motion non-stationary solutions of the system can be obtained. As the differential equations (3.27) and (3.28) are autonomous, the time derivative is eliminated by separation of variables, leading to

$$\frac{d\langle U^2 \rangle}{d\langle N \rangle} = \frac{-\gamma_D \langle U^2 \rangle + \beta \langle U^2 \rangle \langle N \rangle}{\gamma_L \langle N \rangle - \alpha \langle U^2 \rangle \langle N \rangle} . \tag{3.31}$$

This can be rewritten as

$$(\beta - \gamma_D \langle N \rangle^{-1}) \, d\langle N \rangle + (\alpha - \gamma_L \langle U^2 \rangle^{-1}) \, d\langle U^2 \rangle = 0 . \tag{3.32}$$

Integration of equation (3.32) results in the constant of motion. The solution describes oscillatory trajectories around the fix point in phase space. The specific trajectory is determined by the constant on the right hand side:

$$(\beta \langle N \rangle - \gamma_D \ln \langle N \rangle) + (\alpha \langle U^2 \rangle - \gamma_L \ln \langle U^2 \rangle) = \text{const} . \tag{3.33}$$

A representation of the phase space including the stationary solution and some trajectories of the oscillatory solutions is shown in figure 3.6.

For small deviations of the stationary solution (3.30), variables $\langle U^2 \rangle$ and $\langle N \rangle$ can be written as

---

[16]The predator-prey dynamics are also recovered in the cascade (shell) model, a generalisation of the three-wave-interaction model. Additionally, these models exhibit chaotic regimes. [114, 122]

Figure 3.6: Trajectories of a predator-prey model in the phase space, which is spanned by the drift-wave enstrophy $\langle N \rangle$ and the enstrophy of the zonal flow $\langle U^2 \rangle$. The centre of the oscillations marks the stable fix point. [123]

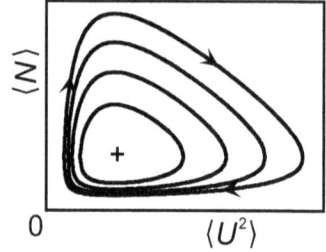

$$\langle N \rangle = \frac{\gamma_D}{\beta} + \delta \langle N \rangle \ , \tag{3.34}$$

$$\langle U^2 \rangle = \frac{\gamma_L}{\alpha} + \delta \langle U^2 \rangle \ . \tag{3.35}$$

Inserting them into the Lotka-Volterra equations (3.27) and (3.28) yields:

$$\frac{\partial}{\partial t} \delta \langle N \rangle = -\gamma_D \frac{\alpha}{\beta} \delta \langle U^2 \rangle - \alpha \, \delta \langle U^2 \rangle \, \delta \langle N \rangle \ , \tag{3.36}$$

$$\frac{\partial}{\partial t} \delta \langle U^2 \rangle = \gamma_L \frac{\beta}{\alpha} \delta \langle N \rangle + \beta \, \delta \langle U^2 \rangle \, \delta \langle N \rangle \ . \tag{3.37}$$

The product $\delta \langle U^2 \rangle \, \delta \langle N \rangle$ is of second order and can be neglected since only small deviations are considered. With this approximation two linear differential equations are obtained,

$$\frac{\partial}{\partial t} \delta \langle N \rangle = -\gamma_D \frac{\alpha}{\beta} \delta \langle U^2 \rangle \ , \tag{3.38}$$

$$\frac{\partial}{\partial t} \delta \langle U^2 \rangle = \gamma_L \frac{\beta}{\alpha} \delta \langle N \rangle \ . \tag{3.39}$$

Differentiation with respect to time leads then to the differential equations of a harmonic oscillator

$$\delta \langle \ddot{N} \rangle + \omega^2 \, \delta \langle N \rangle = \delta \langle \ddot{U}^2 \rangle + \omega^2 \, \delta \langle U^2 \rangle = 0 \ . \tag{3.40}$$

The system has a characteristic frequency of

$$\omega = \sqrt{\gamma_L \gamma_D} \ , \tag{3.41}$$

where the oscillation is only determined by the growth rate and the damping rate of the respective population.

3.3.3 Drift-wave self-regulation

So far the nonlinear self-regulation of the drift waves has been omitted. However, this is essential for drift-wave turbulence and will be accounted for with $\gamma_{\mathrm{NL}} = \gamma_{\mathrm{D2}}\langle N \rangle$. The model equations (3.27) and (3.28) in the extended form are then:

$$\frac{\partial}{\partial t}\langle N \rangle = \gamma_{\mathrm{L}}\langle N \rangle - \gamma_{\mathrm{D2}}\langle N \rangle^2 - \alpha\langle U^2 \rangle\langle N \rangle , \tag{3.42}$$

$$\frac{\partial}{\partial t}\langle U^2 \rangle = -\gamma_{\mathrm{D}}\langle U^2 \rangle + \beta\langle U^2 \rangle\langle N \rangle . \tag{3.43}$$

An additional stationary solution is

$$\langle N \rangle = \frac{\gamma_{\mathrm{L}}}{\gamma_{\mathrm{D2}}} , \quad \langle U^2 \rangle = 0 . \tag{3.44}$$

In this state drift waves exist but without a zonal flow. The growth of the drift waves is thereby determined by the growth rate γ_{L} and, additionally, the nonlinear interaction γ_{D2}. A more interesting solution is

$$\langle N \rangle = \frac{\gamma_{\mathrm{D}}}{\beta} , \tag{3.45}$$

$$\langle U^2 \rangle = \frac{1}{\alpha}\left(\gamma_{\mathrm{L}} - \frac{\gamma_{\mathrm{D2}}\gamma_{\mathrm{D}}}{\beta}\right) . \tag{3.46}$$

Here, also the zonal flow exists. Interestingly, the level of the drift-wave turbulence is only determined by the damping rate of the zonal flow γ_{D}. The solution (3.46) also shows that there exists now a threshold for the growth rate of the drift waves since only positive values for the enstrophy are meaningful,

$$\gamma_{\mathrm{L}} > \frac{\gamma_{\mathrm{D2}}\gamma_{\mathrm{D}}}{\beta} . \tag{3.47}$$

Due to this restriction, the boundaries in phase space are shifted, which leads to a critical minimal growth rate γ_{crit} [124].[17]

When the zonal flow damping is neglected, i.e. $\gamma_{\mathrm{D}} = 0$, the system (3.42) and (3.43) exhibits single burst like events. The phase space portrait of such dynamics is shown on the left hand side of figure 3.7 and the corresponding

---

[17]This corresponds to the Dimits shift regime [125] (collisionless limit, $C \approx 0$) where most of the energy is transferred to the zonal flow and transport values are, therefore, low [126].

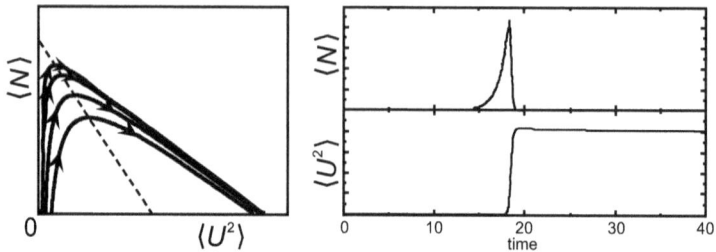

Figure 3.7: Phase space portrait of the predator-prey cycle when the nonlinear self-regulation of the drift waves is included. The corresponding time evolution of the drift-wave enstrophy $\langle N \rangle$ and the zonal flow enstrophy $\langle U^2 \rangle$ is shown on the right hand side exemplary for one trajectory. [123, 128]

time evolution, for further illustration, on the right hand side. In a single burst the complete drift-wave enstrophy is quenched by the zonal flow, which ends up in a steady state.[18]

3.4 Damping mechanisms

Various mechanisms are known to determine the damping of poloidal flows in plasma. In principle, the different damping mechanisms can be grouped into linear and nonlinear effects.

Collisional damping processes belong to the linear damping mechanisms and are due to ion viscosity and resistivity [129]. The influence of friction (collisions between, e.g., ions and neutrals) is mostly disregarded since it is negligible in comparison to the viscosity. But also collisionless damping, like Landau damping, can play a role, as in the case of the GAM.

Also, several nonlinear damping mechanisms, often referred to as saturation mechanisms, can limit the zonal flow growth. As the zonal flow level is ultimately determined by the ambient turbulence, the primary instability (growth and damping rate) essentially defines the saturated state. But the zonal flow can also be unstable to the Kelvin-Helmholtz instability [130], a shear flow instability, which has then the role of a tertiary instability.

In the following, viscous damping will be discussed (Sect. 3.4.1) as an example of collisional damping, and, afterwards (Sect. 3.4.2), the geodesic

---

[18]Such a behaviour has been studied in connection with the LH-transition (cf. Chap. 1). [127]

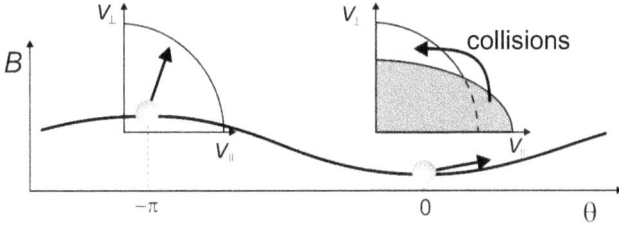

Figure 3.8: Illustration of flow damping due to ion viscosity. In the initial state, on the high field side (left), the velocity distribution, with parallel and perpendicular components, is isotropic. When the particle is moved to the low field side (right), the isotropy of the velocity distribution is broken because energy is transferred between the velocity components (conservation of magnetic moment). Due to collisions, the isotropy can be recovered. This momentum loss leads to the viscous force term in the momentum balance equation. [60]

transfer effect [102] is introduced in more detail, which is the dominant damping mechanism of the zonal flow in toroidal geometry.

3.4.1 Collisional damping

Due to the toroidal geometry of the considered system, neoclassical effects occur which contribute to the damping of the poloidal flow. Such an effect is the ion viscous damping due to magnetic pumping. In any toroidal geometry the magnetic field has to vary in the poloidal or rather parallel direction, why the particles will pass through regions of different magnetic field strength, similar as in a magnetic mirror configuration. The effect on the velocity distribution is sketched in figure 3.8. For the starting position at the high field side the velocity distribution of the particles is assumed to be isotropic. Since the magnetic moment is conserved, the velocity distribution has to adjust when the particles move to a region of different magnetic field strength. Therefore, perpendicular kinetic energy is transferred into parallel kinetic energy, which leads to a deformation of the velocity distribution in phase space. Due to ion-ion collisions, the distribution can regain isotropy. This thermalisation process leads to a momentum lack and manifests itself as a viscous force directed against the poloidal movement. The particles miss momentum to regain their initial state on the high field side.

The neoclassical contributions are included via the viscosity tensor, whose divergence has to be added to the momentum balance equation. Its diagonal elements are connected to the parallel viscosity, which is caused by the magnetic pumping just discussed. The off-diagonal elements are related to gradients in the flow velocity and represent the perpendicular viscosity. Usually, they are disregarded since they do not alter the general behaviour of the viscous damping. The poloidal contribution of the parallel viscosity remains while the toroidal component vanishes because of symmetry and the radial component is comparatively small.[19] In the drive equation (3.17) viscous damping is then just included as

$$\mu_{ii} \frac{\partial^2}{\partial r^2} \bar{v}_\theta \ . \tag{3.48}$$

However, the thermalisation of the velocity distribution crucially depends on the collision frequency of the ions. Surely, without collisions the process of momentum redistribution does not work and the viscous damping is zero (cf. [17]). For increasing collision frequency the viscous damping increases approximately linear. Magnetically trapped particles (banana particles) additionally contribute to the velocity relaxation process. The damping gets maximal when the collision frequency is in the range of the transit frequency of the ions (plateau regime) because the momentum transfer between the velocity components is then maximal. For higher collision frequencies the mean free path decreases and the damping rate gets inversely proportional (Pfirsch-Schlüter regime). When the ion collision frequency is very high, any perturbation in the Maxwell distribution will be quickly thermalised and the velocity distribution is locally in equilibrium.

3.4.2 Geodesic transfer effect

The geodesic transfer mechanism is closely related to the 3D structure of the zonal flow [131]. As described in section 3.1.1, the poloidal flow is not divergence free because of the varying magnetic field, i.e. $\nabla \bar{v}_\theta \propto -\kappa_g/B$. The field line curvature results in $E \times B$-difts directed towards each other for positive geodesic curvature $\kappa_g > 0$, leading to a pressure accumulation $\tilde{p}$ in this region. In a tokamak geometry the resulting density structure is axis-symmetric with poloidal mode number $m = \pm 1$, which is essentially the same spatial structure as the GAM. Via this pressure sidebands the zonal

---

[19]Nevertheless, the secondary (return) flow (see Sect. 3.1.1) is also subject to viscose damping, which has to be added to the damping rate. [102]

flow can couple to the GAM. The major loss channels of the energy in the pressure sidebands is due to diffusive mixing and dissipation involving the adiabatic parallel electron response. In the first case, the energy is nonlinearly transferred from large scales back to smaller scales in the turbulence because of an inhomogeneous pressure advection, i.e. $\mathbf{v}^{E \times B} \nabla \tilde{p}$. The other part of the free energy in the sidebands goes to the parallel electron dynamics. The inhomogeneous pressure perturbation leads to parallel currents $J_{\parallel}$ due to the adiabatic electron response.[20] This is the global Alfvén oscillation, described with the coupling term $\langle J_{\parallel} \cos(s) \rangle$, which is finally dissipated by resistivity of the plasma.

For simple toroidal geometry, similar to tokamaks, the geodesic transfer effect can be modelled by a curvature operator of sinusoidal form. With the additional term the momentum balance equation of the poloidal flow (3.17) reads

$$\frac{\partial}{\partial t} \bar{v}_\theta = -\frac{\partial}{\partial r} \langle \tilde{v}_\theta \tilde{v}_r \rangle - \omega_{\mathrm{B}} \langle p \sin(s) \rangle \ . \tag{3.49}$$

The coordinate s represents the parallel direction in the field aligned flux tube geometry [132], which for a concentric circular geometry is equal to the poloidal angle θ. ω_{B} is the scaling factor of the curvature operator.

In comparison to the other zonal flow damping effects, the geodesic transfer effect is in total the main energy loss of the zonal flow for typical tokamak edge scenarios [113]. Damping due to viscosity or friction is comparably low. Therefore, the magnetic structure plays a crucial role in the damping of the zonal flows and, at the end, for the suppression of the turbulence.

---

[20] In equilibrium, this basically results in the Pfirsch-Schlüter currents, which balance the pressure gradient.

Chapter 4

Data analysis

A statistical description of turbulent systems form the basis of the investigation of turbulence. In this chapter, the basic statistical techniques are introduced in section 4.1–4.3, following [133, 134] and references therein. As the drift-wave zonal-flow coupling is governed by three-wave interaction, bispectral analysis (Sect. 4.4) can be used to investigate the mode coupling. With the solution of the wave-coupling equation also the energy transfer between the turbulent modes can be obtained (Sect. 4.5). Finally, the conditional average is introduced in section 4.6, which is a variation of the ensemble average and can be used to resolve the spatio-temporal dynamics of turbulent structures.

4.1 Basic principles of statistical analysis

In the description of turbulent flows averaged values are used to characterise the overall system state. For an absolute continuous, real valued random variable X with its realisations x, the expected value $E(X)$, or rather the average[1] $\langle\,\cdot\,\rangle$, is defined as

$$\langle X \rangle \equiv E(X) := \int_{-\infty}^{\infty} x P(x)\,\mathrm{d}x \ . \tag{4.1}$$

Convergence of the integral is assumed. The probability density function $P(x)$, defined on $\mathbb{R}$, gives the probability $P(x)\,\mathrm{d}x$ of X lying between x and $x + \mathrm{d}x$. Of course, the density function is normalised, i.e. $\int_{-\infty}^{\infty} P(x)\,\mathrm{d}x = 1$. In the same way as in (4.1), the average of a function solely depending on X is determined with $\langle f(X) \rangle = \int_{-\infty}^{\infty} f(x)P(x)\,\mathrm{d}x$. This can be extended to several random variables[2], here $f(X_1, X_2)$, where the integral is then with

---

[1] The naming and nomenclature is ambiguous and in a mathematical sense not strict. However, the meaning in the specific case throughout this work should be clear.

[2] Fluctuating quantities of a turbulent flow at different times or different points can, for example, be thought of as random variables.

respect to the joint probability density function $P(x_1, x_2)$. For statistically independent variables the joint density function is just the product of the individual ones and it follows that $\langle X_1 X_2 \rangle = \langle X_1 \rangle \langle X_2 \rangle$.

The probability distribution can be described uniquely with the moments of the order k of a random variable X, defined as $m_k(X) := \int_{-\infty}^{\infty} x^k P(x)\, \mathrm{d}x$. For $k = 1$ this corresponds to the mean value. More often, the central moments $\mu_k(X)$ are considered, which are defined relative to the mean with $\tilde{x} = x - \langle X \rangle$,

$$\mu_k(X) := \int_{-\infty}^{\infty} (x - \langle X \rangle)^k P(x)\, \mathrm{d}x = \int_{-\infty}^{\infty} \tilde{x}^k P(x)\, \mathrm{d}x \ . \tag{4.2}$$

In practice, only the first four moments are considered. So the second central moment $\mu_2(X)$, or variance $\mathrm{Var}(X)$, is the expected value of the quadratic deviation to the mean,

$$\mathrm{Var}(X) \equiv \mu_2(X) = \langle \tilde{x}^2 \rangle \ . \tag{4.3}$$

Normally, the square root of the variance is shown, which is called the standard deviation

$$\sigma := \sqrt{\mathrm{Var}(X)} = \sqrt{\mu_2(X)} \ , \tag{4.4}$$

and generally describes the spread of the distribution. For illustration, two normal distributions with different values σ are shown on the left hand side of figure 4.1. The third as well as the fourth central moment is normalised to the standard deviation σ and therefore

$$\mathcal{S} := \frac{\mu_3(X)}{\sigma^3} \ . \tag{4.5}$$

This normalised quantity is the so-called skewness $\mathcal{S}$, which is a measure for the symmetry of the distribution (see middle of Fig. 4.1). Positive values $\mathcal{S} > 0$ indicate a slower decaying wing for higher values than the mean and vice versa for negative skewness $\mathcal{S} < 0$. The kurtosis $\mathcal{K}$, the fourth-order moment, is defined with respect to the normal distribution. This manifests itself as an additional constant in the definition,[3]

$$\mathcal{K} := \frac{\mu_4(X)}{\sigma^4} - 3 \ . \tag{4.6}$$

---

[3] Often the kurtosis is defined without this constant, where then the normal distribution has $\mathcal{K} = 3$.

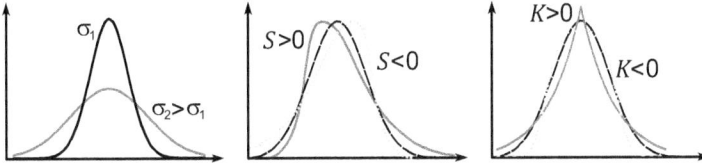

Figure 4.1: Illustration of the standard deviation (left), the skewness (middle) and the kurtosis (right) of a probability distribution function. The black dashed line is a Gaussian distribution.

Figure 4.1 (right) illustrates the meaning of positive and negative values of the kurtosis $\mathcal{K}$. Its importance is due to the instance that it provides information on the decline of the wings of the distribution and, therefore, of rarely occurring large events.[4]

In addition to the moments of a distribution, for jointly distributed random variables X_1 and X_2 the covariance can be defined as

$$\text{Cov}(X_1, X_2) := \langle \tilde{x}_1 \tilde{x}_2 \rangle . \tag{4.7}$$

A normalised quantity is obtained when (4.7) is divided by the standard deviation of both distributions,

$$\rho = \frac{\text{Cov}(X_1, X_2)}{\sigma_1 \sigma_2} . \tag{4.8}$$

This is called the (Pearson) correlation coefficient which measures the linear dependency of the two variables.[5] Its values are bound to the interval $[-1, 1]$, and $\rho = 1$ (correlation) implies a total positive linear relation whereas $\rho = -1$ a negative one (anti-correlation). Independent variables are not correlated and the correlation coefficient is zero since it is $\text{Cov}(X_1, X_2) = \langle \tilde{x}_1 \rangle \langle \tilde{x}_2 \rangle$.[6]

Another way to characterise the distribution is via the characteristic functions, where the first one is introduced as [7]

$$\varphi_X(t) = \langle e^{itX} \rangle = \int_{-\infty}^{\infty} e^{itx} P(x) \, \mathrm{d}x , \tag{4.9}$$

---

[4] Out of this reason it is often used in the study of intermittency [133, 135].

[5] This not the only possible definition of a correlation coefficient, and other measures of a correlation can be sensitive to nonlinear dependencies.

[6] However, $\rho = 0$ does not imply that two random variables are deterministically independent.

[7] The definition of the characteristic functions differs in literature.

which is basically the Fourier transform of the probability density function. With a series expansion of the exponential in equation (4.9), the characteristic function can be written as the sum of the moments of the distribution (4.2),

$$\varphi_X(t) = e^{it\langle X \rangle} \sum_{k=0}^{\infty} \mu_k \frac{(it)^k}{k!} . \tag{4.10}$$

Further on, the second characteristic function is then defined as

$$\psi_X(t) = \log(\varphi_X(t)) , \tag{4.11}$$

and written as a series

$$\psi_X(t) = \sum_{k=1}^{\infty} \kappa_k \frac{(it)^k}{k!} . \tag{4.12}$$

Here, the coefficients κ_k are called the cumulants of X, which are closely connected to the moments (4.2). Analogously, the characteristic functions of joint probability distribution are defined. For an arbitrary number of random variables, the cumulants of the first four orders can be shown to be

$$\begin{aligned}
\kappa_1^{(i)} &= \langle X_i \rangle , \\
\kappa_2^{(ij)} &= \langle \tilde{x}_i \tilde{x}_j \rangle , \\
\kappa_3^{(ijk)} &= \langle \tilde{x}_i \tilde{x}_j \tilde{x}_k \rangle , \\
\kappa_4^{(ijkl)} &= \langle \tilde{x}_i \tilde{x}_j \tilde{x}_k \tilde{x}_l \rangle - \langle \tilde{x}_i \tilde{x}_j \rangle \langle \tilde{x}_k \tilde{x}_l \rangle - \langle \tilde{x}_i \tilde{x}_k \rangle \langle \tilde{x}_j \tilde{x}_l \rangle - \langle \tilde{x}_i \tilde{x}_l \rangle \langle \tilde{x}_j \tilde{x}_k \rangle .
\end{aligned} \tag{4.13}$$

This is especially important for Gaussian distributions since then only the first two cumulants are non-zero, sufficient to completely determine the distribution functions. From $\kappa_4^{(ijkl)} = 0$ then follows

$$\langle \tilde{x}_i \tilde{x}_j \tilde{x}_k \tilde{x}_l \rangle = \langle \tilde{x}_i \tilde{x}_j \rangle \langle \tilde{x}_k \tilde{x}_l \rangle + \langle \tilde{x}_i \tilde{x}_k \rangle \langle \tilde{x}_j \tilde{x}_l \rangle + \langle \tilde{x}_i \tilde{x}_l \rangle \langle \tilde{x}_j \tilde{x}_k \rangle . \tag{4.14}$$

Therefore, the fourth-order moment can be expressed by the sum of second-order moments. This is called the quasi-normal approximation, which will be important for the calculation of the energy transfer (see Sect. 4.5.1).

4.2 Correlation analysis

In turbulence analysis time dependent random variables $X(t)$ are of interest. The covariance can then be defined as $\mathrm{Cov}_{X_1 X_2}(\tau) := \langle \tilde{x}_1(t) \tilde{x}_2(t + \tau) \rangle$, depending on the relative time separation τ. In its normalised form it is known as the cross-correlation function

$$C_{X_1 X_2}(\tau) := \frac{\mathrm{Cov}_{X_1 X_2}(\tau)}{\sigma_1 \sigma_2} . \tag{4.15}$$

Like the correlation coefficient ρ, the correlation function $C_{X_1 X_2}(\tau)$ describes to which degree the relation between X_1 and X_2 can be approximated by a linear relationship. But the correlation function expresses how this dependency changes for a shift of the two signals against each other. A special case is the auto-correlation function $C_{XX}(\tau)$, which, at zero time lag, has the value 1 (complete correlation).

However, the correlation function cannot only be defined in time but also in space, depending on the spatial separation. For a turbulent flow, correlation in space, which is the more natural one, and correlation in time are related and to some degree interchangeable. This is referred to as Taylor hypothesis [136], which is based on the assumption that the convection of the turbulent structures with the mean flow is fast compared to their evolution time. A measurement at a fixed point experiences the turbulent structures moving by as they were frozen in the turbulent flow. Even without a strong mean flow the convection of the small scale structures by the large ones is relatively fast, so that the assumption still holds (at least for a part of the spectrum).

One characteristic of turbulence is that it loses memory of the initial state for longer times or larger distances, respectively. This means that the correlation function has to decay eventually with increasing time or spatial separation, i.e. $C_{X_1 X_2}(\tau) \xrightarrow{\tau \to \infty} 0$ (analogously for spatial separation). Integration over the spatial correlation function results in a length, referred to as correlation length, which is a measure for the largest scales in the turbulence (integral scale, see Chap. 2.1.3). The small scales determine the behaviour of the correlation function for small separations. This already suggests a close relation between the correlation function and the cascades, which will become apparent in the next section.

Of course, the cross-correlation function can also be examined for time signals at separated spatial locations. When a fixed reference signal is chosen, the spatio-temporal evolution of quasi-coherent structures can be obtained [137]. This can lead to similar results as the conditional averaging technique presented in section 4.6 [103, 138].

4.3 Spectral analysis

4.3.1 Fourier transformation

Only discrete time traces can be examined in practice. The Fourier transform of a time trace with N time points t_n with $n \in [0, N-1]$ is defined as

$$\widehat{X}(\omega_m) \equiv \mathcal{F}_{\omega_m}(X(t)) := \frac{1}{N} \sum_{n=0}^{N-1} x(t_n) e^{-i\omega_m t_n} . \tag{4.16}$$

An equal spacing δt is assumed, which corresponds to the frequency resolution of $\delta\omega = 2\pi/(N\,\delta t)$. The complex valued Fourier coefficients $\widehat{X}(\omega_m)$ are associated with $\omega_m = 2\pi f_m = m\,\delta\omega$ with $m \in [0, N-1]$.

The cross-spectrum is given as the product of two Fourier transforms,

$$H_{X_1 X_2}(\omega) := \langle \widehat{X}_1^*(\omega)\widehat{X}_2(\omega) \rangle = h_{X_1 X_2}(\omega) e^{i\delta_{X_1 X_2}(\omega)} , \tag{4.17}$$

and, via the Wiener-Khintchin theorem, related to the cross-correlation, i.e. $H_{X_1 X_2}(\omega) = \mathcal{F}(C_{X_1 X_2}(\tau))$. The cross-spectrum measures to which extent the system can be described by a linear functional dependency. If $X_1 = X_2$, then the result is real-valued and called auto (power) spectrum $S_X(\omega)$, representing the spectral power density in the signal.[8] In general, the cross-spectrum is a complex quantity and may be expressed with the cross-amplitude spectrum $h_{X_1 X_2}(\omega)$ and the cross-phase spectrum $\delta_{X_1 X_2}(\omega)$. Normalised to the respective auto-spectra, this results in the cross-coherence spectrum

$$\gamma_{X_1 X_2}(\omega) = \frac{h_{X_1 X_2}(\omega)}{\sqrt{S_{X_1 X_1}(\omega) S_{X_2 X_2}(\omega)}} , \tag{4.18}$$

which is then restricted to the interval $[0, 1]$. For a constant phase relation the coherence does not vanish. When the coherence has substantial values, the cross-phase spectrum is meaningful and can be calculated from imaginary and real part of the cross-spectrum,

$$\delta_{X_1 X_2}(\omega) = \arctan\left(\frac{\mathrm{Im}[H_{X_1 X_2}(\omega)]}{\mathrm{Re}[H_{X_1 X_2}(\omega)]}\right) . \tag{4.19}$$

---

[8]This shows the before mentioned close connection between the correlation function and the turbulent cascade as $S_X(\omega) = \mathcal{F}(\mathrm{Cov}_{XX}(\tau))$.

The Fourier transform (4.16) is, in principle, not restricted to time series $X(t)$ but can also be applied to spatial data fields $X(s)$. The spectrum of the spatial scales $S(k_m)$ is then a function of the wavevector $k_m = 2\pi/\lambda_m = m\,\delta k$, with the wavelength λ_m. Analogue to the connection between time and frequency, the spatial separation δs determines the resolution in k-space which is $\delta k = 2\pi/(N\,\delta s)$.

When the temporal evolution of a whole data field can be measured ($X = X(t, s)$), then the joint spectrum in frequency and wavenumber space can be calculated, called kf-spectrum

$$S(k, f) := \mathcal{F}_k(\mathcal{F}_f(X(t, s))) \,. \tag{4.20}$$

This gives information about the frequency for a certain spatial scale and shows, therefore, directly the dispersion relation (see Sect. 6.3.2). To cover the statistical average, the ensemble average can be taken as average over subseries of the whole time series which do not overlap. The resolution is then limited by the length of the subseries.

4.3.2 Wavelet transformation

The wavelet transformation can be seen as an expansion of the Fourier transform where the signal is decomposed into scales (potentially interpreted as frequencies) and time. For a time trace of N discrete time points $t_n = n\,\delta t$ it is defined as

$$\mathcal{W}_{m,a}(X(t)) := \sum_{n=0}^{N-1} x(n\,\delta t)\,\Psi_a^*\left((n-m)\,\delta t\right) \,, \tag{4.21}$$

with $n, m \in [0, N-1]$, specifying the time point t_m for which the wavelet coefficient $\mathcal{W}_{m,a}$ is calculated. By this the signal is bandpass filtered for the frequency corresponding to the wavelet scale a. In here, the Morlet wavelet is used as kernel function of the wavelet transformation [139], i.e.

$$\Psi_a(t) = \frac{1}{\sqrt{a}}\pi^{-\frac{1}{4}}\exp\left[i\frac{2\pi t}{a} - \frac{1}{2}\left(\frac{t}{a}\right)^2\right] \,. \tag{4.22}$$

It is composed of a sinusoidal oscillation and a Gaussian decay. For this wavelet the scale is connected to the frequency according to $\omega = (2\pi + \sqrt{2 + 4\pi^2})/2a$. Similar as in the case of the Fourier transform, the wavelet transformation can be transferred to spatial data. Using the wavelet transformation, the cross-spectrum as well as higher-order spectra may thus be generalised to keep a temporal or spatial resolution [140–142].

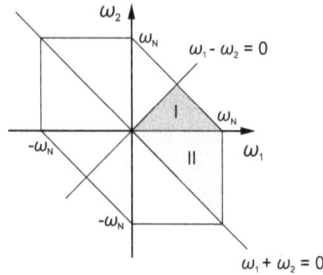

Figure 4.2: Domain of definition for the bispectrum of discrete time traces. The Nyquist frequency $\pm\omega_N$ is the limit for the frequencies ω_1, ω_2 and also for the sum $\omega_1 + \omega_2$. The diagonal and counter diagonal are distinguished with the requirement $\omega_1 - \omega_2 = 0$ and $\omega_1 + \omega_2 = 0$, respectively. Because of symmetry to these axes, the domain is restricted to the areas I and II. Furthermore, region II can be mapped onto region I. [143]

4.4 Bispectral analysis

In principle, the definition of the cross-spectrum (4.17) can be extended to an arbitrary number of factors. But especially for turbulence, the Fourier transform of the third-order cumulant (cf. Eq. (4.13)) is of major interest. Important interaction mechanisms, like between drift waves and zonal flows, are governed by three-wave interaction (see Chap. 3.2) and, therefore, the bispectrum is an often used quantity. The cross-bispectrum is the triple product of the Fourier transformations of the signals X_1, X_2, and X, depending on the frequencies ω_1 and ω_2,

$$B(\omega_1, \omega_2) := \langle \widehat{X}_1(\omega_1)\widehat{X}_2(\omega_2)\widehat{X}^*(\omega_1 + \omega_2) \rangle . \qquad (4.23)$$

The components have to fulfil the resonance condition $\omega = \omega_1 + \omega_2$. When the spectrum is calculated from a single signal it is called the auto-bispectrum, and, also here, it is related to the Fourier transform of the triple correlation function $C(\tau_1, \tau_2)$. The definition (4.23) is easily generalised for spatial data with the Fourier transformation in k-space.

For discrete time traces the domain of definition is restricted to the hexagon shown in figure 4.2. The frequencies ω_1, ω_2, and $\omega = \omega_1 + \omega_2$ are limited to frequencies below the Nyquist frequency $\pm\omega_N$. Since the Fourier transform is Hermitian, the bispectrum is subject to the following symmetries [144],

$$B(\omega_1, \omega_2) = B(\omega_2, \omega_1) = B^*(-\omega_1, -\omega_2) \qquad (4.24)$$

$$= B^*(-\omega_1, \omega_1 + \omega_2) = B^*(-\omega_2, \omega_1 + \omega_2) . \qquad (4.25)$$

With (4.24) follows the symmetry to the diagonal $\omega_1 - \omega_2 = 0$ and counter diagonal $\omega_1 + \omega_2 = 0$, which leaves region I and II in the figure. Furthermore, with equation (4.25) region II is mapped onto region I which is then the only non-redundant domain.

As in the case of the cross-spectrum, it is desired to obtain a quantitative measure of the nonlinear coupling between the three components [69]. The square of the absolute value of the bispectrum is usually normalised to the cross-spectrum and the auto-spectrum which gives then the quadratic bicoherence

$$b^2(\omega_1, \omega_2) := \frac{|\langle \widehat{X}_1(\omega_1)\widehat{X}_2(\omega_2)\widehat{X}^*(\omega_1 + \omega_2)\rangle|^2}{\langle|\widehat{X}_1(\omega_1)\widehat{X}_2(\omega_2)|^2\rangle\langle|\widehat{X}(\omega_1 + \omega_2)|^2\rangle} . \qquad (4.26)$$

With the normalisation[9] the bicoherence is limited to the range $[0, 1]$ and, since (4.25) does not hold, has region I and II as its domain of definition. For statistically independent modes, with their individual phase θ, the biphase

$$\beta(\omega_1, \omega_2) := \theta(\omega_1) + \theta(\omega_2) - \theta(\omega_1 + \omega_2) , \qquad (4.27)$$

is equally distributed on the interval $(-\pi, \pi]$ and, therefore, the bispectrum and bicoherence is averaged out. When the components are quadratically phase coupled, the bicoherence does not vanish and determines the coupling strength.[10]

4.5 Analysis of the energy transfer

The bispectrum (Sect. 4.4) provides a measure of a mode coupling but it does not show in which direction the energy is transferred. One way to calculate the energy transfer in the turbulence is to solve the wave-coupling equation [147]

$$\frac{\partial \varphi(k, t)}{\partial t} = \Lambda_k^L(k)\varphi(k, t) + \frac{1}{2} \sum_{\substack{k_1, k_2 \\ k = k_1 + k_2}} \Lambda_k^Q(k_1, k_2)\varphi(k_1, t)\varphi(k_2, t) . \quad (4.28)$$

---

[9]Other normalisations can be chosen which then are not necessarily restricted to a maximal value of 1 [145].

[10]Spurious values of the bicoherence can occur due to decayed nonlinear processes or nonlinear measurement effects [146].

The evolution of the fluctuating quantity $\varphi(k,t)$, here defined in k-space but transferable to frequency space, is determined by a linear part and a nonlinear part. The linear part is build by the growth rate γ_k and the dispersion ω_k of the wave, included in the linear transfer function $\Lambda_k^L(k) = \gamma_k + i\omega_k$. The nonlinear part represents the coupling of the different modes, which obey $k = k_1 + k_2$, where the coupling strength is specified by the nonlinear, or quadratic, coupling coefficient $\Lambda_k^Q(k_1, k_2)$. For the following approach the fluctuating quantity is written with the complex exponential function, i.e. $\varphi(k,t) = |\varphi(k,t)|e^{i\Theta(k,t)}$. With this the temporal derivative can be expressed by the difference quotient as

$$\frac{\partial \varphi(k,t)}{\partial t} = \lim_{\tau \to 0}\left(\frac{|\varphi(k,t+\tau)| - |\varphi(k,t)|}{\tau} \frac{1}{|\varphi(k,t)|} \right.$$
$$\left. + i\frac{\Theta(k,t+\tau) - \Theta(k,t)}{\tau} \right) \cdot \varphi(k,t) \ . \quad (4.29)$$

For small time differences τ compared to the evolution of a wave package, the time derivative in the wave-coupling equation (4.28) can be replaced by this difference quotient. After some simplifications and rearrangement, a time discrete version of the wave-coupling equation is obtained,[11]

$$Y_k = L_k X_k + \frac{1}{2} \sum_{\substack{k_1,k_2 \\ k=k_1+k_2}} Q_k^{k_1,k_2} X_{k_1} X_{k_2} \ . \quad (4.30)$$

Input signal and output signal are now

$$X_k = \varphi(k,t) \ , \qquad Y_k = \varphi(k,t+\tau) \ . \quad (4.31)$$

As in the wave-coupling equation (4.28), the transfer functions keep their role but include an additional phase factor. They are now defined for a discrete time separation where the (discrete) linear transfer function reads

$$L_k = \frac{\Lambda_k^L \tau + 1 - i[\Theta(k,t+\tau) - \Theta(k,t)]}{e^{-i[\Theta(k,t+\tau)-\Theta(k,t)]}} \ . \quad (4.32)$$

And the (discrete) nonlinear transfer function is

$$Q_k^{k_1,k_2} = \frac{\Lambda_k^Q(k_1,k_2)\tau}{e^{-i[\Theta(k,t+\tau)-\Theta(k,t)]}} \ . \quad (4.33)$$

---

[11]The factor $1/2$ is omitted when the summation is restricted to $k_1 \geq k_2$.

The transfer functions determine the energy transfer in the turbulence, which becomes clear when the evolution of the spectral power $P_k = \langle \varphi(k,t)\varphi^*(k,t)\rangle$ is considered,

$$\frac{\partial}{\partial t}(\varphi(k,t)\varphi^*(k,t)) = \frac{\partial \varphi(k,t)}{\partial t}\varphi^*(k,t) + \varphi(k,t)\frac{\partial \varphi^*(k,t)}{\partial t} . \tag{4.34}$$

When the wave-coupling equation (4.28) is inserted into equation (4.34), the wave kinetic equation is derived,

$$\frac{\partial}{\partial t}P_k = 2\gamma_k P_k + \sum_{\substack{k_1,k_2 \\ k=k_1+k_2}} T_k(k_1,k_2) . \tag{4.35}$$

In a stationary state, the linear mechanisms, i.e. growth rate and dispersion, are balanced by the nonlinear transfer effects. Especially for zonal flows, which are nonlinearly driven by the turbulence, the nonlinear spectral power transfer function $T_k(k_1,k_2)$ is of interest for the analysis, which is given as

$$T_k(k_1,k_2) = \text{Re}(\Lambda_k^Q(k_1,k_2) \langle \varphi(k_1,t)\varphi(k_2,t)\varphi^*(k,t)\rangle) . \tag{4.36}$$

4.5.1 Ritz method

To obtain the transfer functions L_k and $Q_k^{k_1,k_2}$ the wave-coupling equation (4.30) is transformed into a set of moment equations. This represents an equivalent problem to the equation of motion and also includes a statistical average. For this purpose, equation (4.30) is multiplied by X_k^* and averaged which results in

$$\langle Y_k X_k^*\rangle = L_k \langle X_k X_k^*\rangle + \frac{1}{2} \sum_{\substack{k_1,k_2 \\ k=k_1+k_2}} Q_k^{k_1,k_2} \langle X_{k_1} X_{k_2} X_k^*\rangle . \tag{4.37}$$

Just as well, a multiplication of equation (4.30) by $X_{k_1'}^* X_{k_2'}^*$ and a subsequent average leads to

$$\langle Y_k X_{k_1'}^* X_{k_2'}^*\rangle = L_k \langle X_k X_{k_1'}^* X_{k_2'}^*\rangle + \frac{1}{2} \sum_{\substack{k_1,k_2 \\ k=k_1+k_2}} Q_k^{k_1,k_2} \langle X_{k_1} X_{k_2} X_{k_1'}^* X_{k_2'}^*\rangle , \tag{4.38}$$

where $k = k_1 + k_2 = k_1' + k_2'$ holds.[12] The procedure, which led to the momentum equations (4.37) and (4.38), could be repeated on and on, resulting

---

[12] For both equations, the transfer functions were pulled trough the average although they include the phase difference. As in [148], the phase is approximated as the averaged cross-phase spectrum.

in an infinit set of momentum equations (closure problem, Chap. 1.1). At this place, the Ritz method [149] assumes that the fluctuating quantities have Gaussian or rather near Gaussian statistics [150]. Then, the fourth-order moment can be approximated with the sum of second-order moments (Eq. (4.14)) and $\langle X_{k_1} X_{k_2} X_{k_1'}^* X_{k_2'}^* \rangle$ is replaced by $\langle |X_{k_1} X_{k_2}|^2 \rangle$.[13] This leads to the simplified version of equation (4.38), which now includes only terms of third order,

$$\langle Y_k X_{k_1}^* X_{k_2}^* \rangle \approx L_k \langle X_k X_{k_1}^* X_{k_2}^* \rangle + \frac{1}{2} Q_k^{k_1, k_2} \langle |X_{k_1} X_{k_2}|^2 \rangle \,. \tag{4.39}$$

With the quasi-normal approximation, the hierarchy of equations is closed and can now be solved for the discrete transfer functions. Rearranging equation (4.39) produces the quadratic transfer function

$$Q_k^{k_1, k_2} = 2 \frac{\langle Y_k X_{k_1}^* X_{k_2}^* \rangle - L_k \langle X_k X_{k_1}^* X_{k_2}^* \rangle}{\langle |X_{k_1} X_{k_2}|^2 \rangle} \,. \tag{4.40}$$

And together with equation (4.37) this leads to the linear transfer function

$$L_k = \frac{\langle Y_k X_k^* \rangle - \sum_{\substack{k_1, k_2 \\ k = k_1 + k_2}} \frac{\langle Y_k X_{k_1}^* X_{k_2}^* \rangle \langle X_k^* X_{k_1} X_{k_2} \rangle}{\langle |X_{k_1} X_{k_2}|^2 \rangle}}{\langle X_k X_k^* \rangle - \sum_{\substack{k_1, k_2 \\ k = k_1 + k_2}} \frac{\langle X_k X_{k_1}^* X_{k_2}^* \rangle \langle X_k^* X_{k_1} X_{k_2} \rangle}{\langle |X_{k_1} X_{k_2}|^2 \rangle}} \,. \tag{4.41}$$

Finally, the linear and nonlinear transfer functions, (4.32) and (4.33), and the spectral power transfer function (4.36) can be calculated from the auto-power spectrum $\langle X_k X_k^* \rangle$, the cross-power spectrum $\langle Y_k X_k^* \rangle$, the auto-bispectrum $\langle X_k^* X_{k_1} X_{k_2} \rangle$, and the cross-bispectrum $\langle Y_k X_{k_1}^* X_{k_2}^* \rangle$.

4.5.2 Kim method

In contrast to the Ritz method, now the full fourth-order moments will be retained for the calculation of the energy transfer. The Kim method [153] enforces the stationary condition of the spectrum,

$$\frac{\partial}{\partial t} P_k \approx \frac{\langle Y_k Y_k^* \rangle - \langle X_k X_k^* \rangle}{\tau} = 0 \,, \tag{4.42}$$

in order to obtain a complete set of equations. Equation (4.42) results in

$$\langle Y_k Y_k^* \rangle = \langle X_k X_k^* \rangle \,. \tag{4.43}$$

---

[13] The additional terms with $(k_1', k_2') \neq (k_1, k_2)$ can be neglected [151, 152].

Another equation for the spectral power $\langle Y_k Y_k^* \rangle$ is obtained when equation (4.30) is multiplied with the conjugate Y_k^* and averaged,

$$\langle Y_k Y_k^* \rangle = L_k \langle X_k Y_k^* \rangle + \frac{1}{2} \sum_{\substack{k_1,k_2 \\ k=k_1+k_2}} Q_k^{k_1,k_2} \langle X_{k_1} X_{k_2} Y_k^* \rangle . \qquad (4.44)$$

With equation (4.37) and (4.38) there are now four equations for the four unknown variables L_k, $Q_k^{k_1,k_2}$, $\langle Y_k Y_k^* \rangle$, and $\langle X_k X_k^* \rangle$. For the further procedure, the system of equations is written in matrix notation. The corresponding vectors and matrices are defined, for an even mode number l, as follows:

$$\mathbf{Q} = \left(Q_l^{\frac{l}{2},\frac{l}{2}}, Q_l^{\frac{l+2}{2},\frac{l-2}{2}}, Q_l^{\frac{l+4}{2},\frac{l-4}{2}}, \ldots, Q_l^{l_{\mathrm{N}},l-l_{\mathrm{N}}} \right) , \qquad (4.45)$$

$$\mathbf{A} = \left(\langle X_{\frac{l}{2}} X_{\frac{l}{2}} X_l^* \rangle, \langle X_{\frac{l-2}{2}} X_{\frac{l+2}{2}} X_l^* \rangle, \langle X_{\frac{l-4}{2}} X_{\frac{l+4}{2}} X_l^* \rangle, \ldots \right.$$
$$\left. \ldots, \langle X_{l_{\mathrm{N}}} X_{l-l_{\mathrm{N}}} X_l^* \rangle \right)^T , \qquad (4.46)$$

$$\mathbf{B} = \left(\langle X_{\frac{l}{2}} X_{\frac{l}{2}} Y_l^* \rangle, \langle X_{\frac{l-2}{2}} X_{\frac{l+2}{2}} Y_l^* \rangle, \langle X_{\frac{l-4}{2}} X_{\frac{l+4}{2}} Y_l^* \rangle, \ldots \right.$$
$$\left. \ldots, \langle X_{l_{\mathrm{N}}} X_{l-l_{\mathrm{N}}} Y_l^* \rangle \right)^T , \qquad (4.47)$$

$$\mathbf{F} = \begin{pmatrix} \langle X_{\frac{l}{2}} X_{\frac{l}{2}} X_{\frac{l}{2}}^* X_{\frac{l}{2}}^* \rangle & \langle X_{\frac{l}{2}} X_{\frac{l}{2}} X_{\frac{l+2}{2}}^* X_{\frac{l-2}{2}}^* \rangle & \\ \langle X_{\frac{l+2}{2}} X_{\frac{l-2}{2}} X_{\frac{l}{2}}^* X_{\frac{l}{2}}^* \rangle & \langle X_{\frac{l+2}{2}} X_{\frac{l-2}{2}} X_{\frac{l+2}{2}}^* X_{\frac{l-2}{2}}^* \rangle & \vdots \\ & \ldots & \langle X_{l_{\mathrm{N}}} X_{l-l_{\mathrm{N}}} X_{l_{\mathrm{N}}}^* X_{l-l_{\mathrm{N}}}^* \rangle \end{pmatrix} . $$
$$(4.48)$$

The indices run up to the index l_{N} limited by the corresponding Nyquist frequency. For odd mode numbers the counting index l, in the first two components of each entry (all in the case of $\mathbf{F}$), is replaced by $(l \pm 1)$, respectively. In the matrix notation, the set of equations (4.37), (4.38), (4.43), and (4.44) takes the following form:

$$\langle Y_k X_k^* \rangle = L_k \langle X_k X_k^* \rangle + \mathbf{Q} \cdot \mathbf{A} . \qquad (4.49)$$

$$\langle Y_k Y_k^* \rangle = L_k \langle X_k Y_k^* \rangle + \mathbf{Q} \cdot \mathbf{B} . \qquad (4.50)$$

$$(\mathbf{B}^*)^T = L_k (\mathbf{A}^*)^T + \mathbf{Q} \cdot \mathbf{F} . \qquad (4.51)$$

$$\langle Y_k Y_k^* \rangle = \langle X_k X_k^* \rangle . \qquad (4.52)$$

From here on it is easy to solve for the transfer functions L_k and $\mathbf{Q}$. First, equation (4.51) yields the quadratic transfer function

$$\mathbf{Q} = (\mathbf{B}^*)^T \cdot \mathbf{F}^{-1} - L_k (\mathbf{A}^*)^T \cdot \mathbf{F}^{-1} \; . \tag{4.53}$$

Then, by either substituting this into equation (4.49) or (4.50), the linear transfer function is obtained,

$$L_k = \frac{\langle Y_k X_k^* \rangle - (\mathbf{B}^*)^T \cdot \mathbf{F}^{-1} \cdot \mathbf{A}}{\langle X_k X_k^* \rangle - (\mathbf{A}^*)^T \cdot \mathbf{F}^{-1} \cdot \mathbf{A}} \; , \tag{4.54}$$

or

$$L_k = \frac{\langle Y_k Y_k^* \rangle - (\mathbf{B}^*)^T \cdot \mathbf{F}^{-1} \cdot \mathbf{B}}{\langle X_k Y_k^* \rangle - (\mathbf{A}^*)^T \cdot \mathbf{F}^{-1} \cdot \mathbf{B}} \; . \tag{4.55}$$

Similar as in section 4.5.1, the linear and nonlinear transfer functions, Λ_k^L and $\Lambda_k^Q(k_1, k_2)$, can now be calculated. In chapter 9 this method is used to calculate the energy transfer between the drift waves and the zonal flow.

4.6 Conditional averaging

Up to now the average was assumed to be taken over all available realisations, but often the average of a variable X_1 during a certain state of the system is desired. When a condition for this state, indicated by variable X_2, can be formulated, the average can be restricted to the realisations where the condition is fulfiled. There is no constraint on the specific condition, however the variable is often chosen to be above a critical value, $X_2 \geq x_{\mathrm{crit}}$. The average of X_1 conditioned on X_2 is written as

$$\langle X_1 | X_2 \rangle := \int_{-\infty}^{\infty} x_1 P(x_1 | x_2) \, \mathrm{d}x_1 \; . \tag{4.56}$$

As for the expected value (Def. (4.1)), $P(x_1 | x_2) \, \mathrm{d}x_1$ is the probability for the occurrence of variable X_1 in the interval $[x_1, x_1 + \mathrm{d}x_1]$ but now within the subensemble. With the restriction to a subensemble, the number of realisations decreases why often longer time series are needed for converged statistics.

The definition for the conditional average can be extended to time dependent functions. When the condition can be narrowed down to specify single time points t_i, then also the averaged time evolution around the trigger condition can be obtained. For the N trigger points t_i, time windows are extracted from the signal X_1 as $[t_i - T/2, t_i + T/2]$. Here, T is the interval length, which should be large enough to capture the dynamics and short enough so that the subwindows do not overlap. The average is a function of the relative time lag $\tau \in [-T/2, T/2]$ and is calculated as $\langle X_1 | X_2 \rangle (\tau) = 1/N \sum_{i=1}^{N} x_1(t_i + \tau)$.

Eventually, this can be further extended to time signals at multiple positions, where the spatio-temporal evolution conditioned on a specific event occurring in space and time can be obtained [154–156]. The conditional average is then, additionally, a function of the spatial separation Δr to the reference signal X_{ref} at a fixed position r_{ref}, i.e.

$$\langle X | X_{\mathrm{ref}} \rangle (\Delta r, \tau) = \frac{1}{N} \sum_{i=1}^{N} x(r_{\mathrm{ref}} + \Delta r, t_i + \tau) \, . \qquad (4.57)$$

$$\tau \in [-T/2, T/2]$$

In this procedure, T should also be large enough to capture a possible spatial propagation. The result is an averaged signal at the position of the test signal $r_{\mathrm{ref}} + \Delta r$ with respect to the trigger event at the reference position r_{ref}. With a spatially fixed reference and a moving probe, obtaining the test signal, it is possible to determine the averaged time evolution in larger areas. This is, e.g., used to obtain the zonal flow dynamics in the poloidal cross section shown in chapter 7.2.

Chapter 5

Experiment and diagnostics

Stellarators are best suited to maintain a toroidal low temperature plasma at stationary conditions. The experiment TJ-K, where the measurements of this work have been conducted, is introduced in section 5.1. Thereafter, the working principle of the Langmuir-probe diagnostic is explained (Sect. 5.2.1), and the application of array configurations to measure complex flow quantities is shown (Sects. 5.2.2 and 5.2.3).

5.1 The Stellarator TJ-K

Originally built at CIEMAT in Madrid, Spain, the stellarator experiment TJ-K is now located at the University of Stuttgart in Germany. The experiment is of type torsatron, where the toroidal magnetic field is generated by a single helical field coil ($l = 1$), with 120 windings, going six times poloidally around the vacuum vessel ($m = 6$). The Helmholtz like pair of coils, with 93 windings each, compensates the vertical field component to obtain closed flux surfaces inside the vacuum vessel. Due to a poloidal field component, the field lines are twisted around the torus as to compensate local charge accumulations along the field line. The experiment is operated with a current up to 2 kA, which corresponds to a magnetic field strength of roughly 500 mT on axis. Currently, three microwave heating systems at frequencies 2.45 GHz, 8 GHz, and 14 GHz are available with a fourth (at 28 GHz) ready to be installed. A variety of gases can be used in the experiment with ion masses ranging from $m_i^{\mathrm{H}} \approx 1\,\mathrm{u}$ up to $m_i^{\mathrm{Kr}} \approx 84\,\mathrm{u}$. With all these parameter adjustments at hand, a broad parameter space can be accessed. Typical parameters are summarised in table 5.1. Although the achieved parameters are comparatively low, it has been shown that normalised quantities are similar to those in fusion edge plasmas [157, 158]. Furthermore, many studies demonstrated the drift-wave nature of the plasma turbulence in TJ-K plasmas with a density-potential cross phase close to

Major plasma radius	$R_0 = 0.6\,\text{m}$
Minor plasma radius	$a = 0.1\,\text{m}$
Magnetic field	$B = 50\,\text{mT} - 300\,\text{mT}$
Microwave heating	3 kW at 2.45 GHz
	2.4 kW at 8 GHz
Pulse duration	up to 45 min
Gas	H_2, D_2, He, Ne, Ar, Kr
Electron temperature	$T_e \approx 10\,\text{eV}$
Ion temperature	$T_i \leq 1\,\text{eV}$
Electron density	$n_e = 1 \cdot 10^{18}\,\text{m}^{-3}$
Rotational transform	$\iota = 0.13 - 0.4$

Table 5.1: Typical parameters of the experiment TJ-K.

zero and finite parallel wavelength [88–91]. The magnetic component $\tilde{B}$ in the turbulence is negligible, with only a small Alfvénic contribution to the parallel dynamic. Especially for small ion masses the ρ_s scaling was found to be close to predictions for drift-wave turbulence [159]. Biasing experiments result in enhanced long-range correlations [36, 160–162] and increased zonal flow power proportional to the bias voltage [163].

5.1.1 Experimental setup

The 24 access ports of the experiment guarantee excellent accessibility. They are located at the bottom (B), top (T), inside (I), and outside (O) of the torus, dividing the experiments ideally in six identical segments. But, due to small errors in the magnetic field [164, 165], the toroidal symmetry is broken, and the flux surfaces at each port with similar position are of slightly different shape. The numbering is shown on the schematic view in figure 5.1.

For this work two poloidal limiters were built, which were used for all the experiments herein. The limiters are specifically designed for the ports O3 and O5 with a large plasma cross section compared to the ones used previously in [166, 167]. They define the separatrix at the inside and outside on the midplane ($z = 0\,\text{cm}$) at $R - R_0 = -4.8\,\text{cm}$ and 11.6 cm (port O3) and $R - R_0 = -5.0\,\text{cm}$ and 12.2 cm (port O5).[1] Therefore, the separatrix is everywhere well-defined and the connection length is homogeneous in the scrape-off layer (SOL).

---

[1] This corresponds to the starting position $z = 0\,\text{cm}$ and $R_{\text{start}} = 63.85\,\text{cm}$ at the toroidal position $\varphi = 0°$ (port I1), conventionally used in the field line tracing code [168].

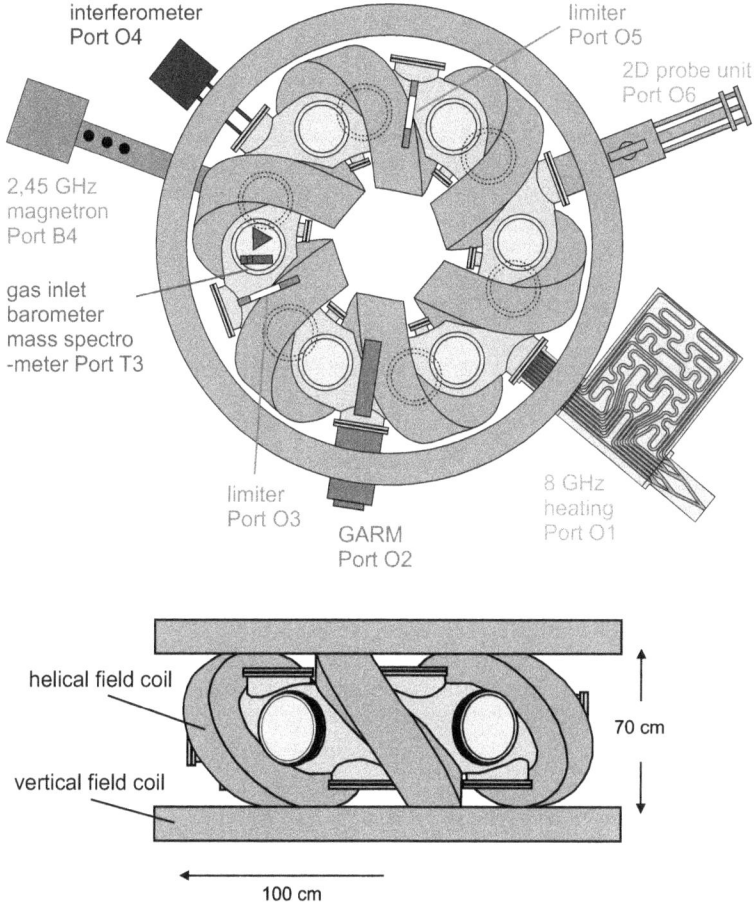

interferometer
Port O4

limiter
Port O5

2D probe unit
Port O6

2,45 GHz
magnetron
Port B4

gas inlet
barometer
mass spectro
-meter Port T3

limiter
Port O3

GARM
Port O2

8 GHz
heating
Port O1

helical field coil

vertical field coil

70 cm

100 cm

Figure 5.1: Schematic overview of the experiment TJ-K in top and side view.

Discharges with 2.45 GHz and 8 GHz microwave heating are analysed. The 2.45 GHz magnetron is positioned at port B4 and the klystron, with output frequencies between 7.9 GHz and 8.4 GHz, at port O1. With a phased array antenna, the different frequencies result in a tilt of the emitted beam, where a frequency of 8.256 GHz is used for optimal absorption [169]. A microwave heating power of up to 3 kW and 2.4 kW, respectively, is available. It was shown [170–172] that the plasma is ignited at the electron cyclotron resonance and then, when the density is high enough, maintained at the upper hybrid resonance, both located in the edge of the confined region. In the case of 8 GHz heating, additionally, an O-X-B mode conversion process takes place and an electrostatic electron Bernstein wave is excited, which is not limited by density cutoffs and can be efficiently absorbed near the plasma centre. Generally, higher heating frequencies require higher magnetic fields and, in the case of 8 GHz heating, the shot duration is limited to a few minutes.

Several standard diagnostics are available to measure plasma parameters. The pressure and the gas composition are monitored with a pressure gauge (Pirani and Penning sensor) down to 0.1 mPa and a mass spectrometer at port T3. At port O4 a microwave interferometer is installed, measuring the line-averaged density, which is used to normalise density profiles from radial probe measurements. With the probe diagnostics at port O6 also electron temperature and plasma potential profiles are available (see Sect. 5.2.1). The knowledge of these parameters is crucial to conduct scaling analysis. But also the temporal evolution of the turbulence can be studied with this 2D-movable probe unit, further described in section 5.2.2. Due to the good accessibility and the low temperature plasmas, probe arrays are often used to measure turbulent fluctuations at multiple points at the same time. For the analysis of the edge turbulence, especially the zonal flow, the poloidal Reynolds stress array (Sect. 5.2.3) is used at port O2.

5.1.2 Magnetic field structure

The shape of the flux surfaces and the position of the magnetic axis[2] is determined by the ratio of the currents in the vertical and helical field coils, $r_{v/h} = I_v/I_h$. Nested flux surfaces inside the vacuum vessel are obtained for ratios between $r_{v/h} = 49\%$ and 65%, where the standard current ratio is

---

[2]Magnetic field parameters can be calculated with a field line tracing code [168] based on the Gourdon code [173].

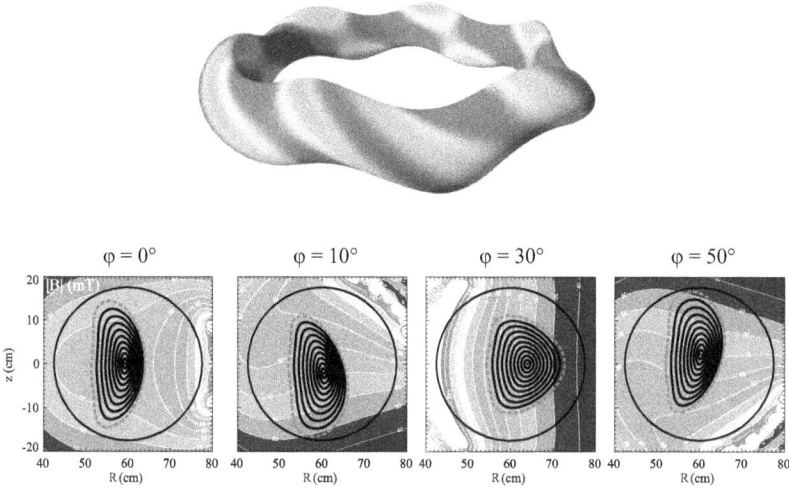

Figure 5.2: Visualisation of the magnetic field geometry of the stellarator experiment TJ-K for standard current ratio $r_{v/h} = 57\%$. *Bottom:* Corresponding cross sections at the different types of access ports. The magnetic field strength is calculated for a value of $I_h = 300$ A. Flux surfaces in the confined region are shown as solid black lines whereas in the scrape-off layer (SOL) as dashed red. Figures from reference [101].

at 57 %.[3] For the values 53 %, 56 %, 58.5 %, and 60 % magnetic islands with increasing mode number arise inside the confined region. A visualisation of a magnetic flux surface, onto which the magnetic field strength is colour coded, is shown in figure 5.2. Below are the cross sections at the different types of access ports for a helical field current of $I_h = 300$ A. The magnetic field strength varies strongly on the flux surface whereupon the in-out variation due to toroidicity is overlaid with the band like structure following the helical field coil. The toroidal periodicity ($m = 6$) transfers to the sixfold symmetry in the magnetic field. With the toroidal angle the position of the magnetic

---

[3] The current ratio for the field line tracing is slightly different where 56.6 % in the code corresponds to 57 % in the experiment. The numbers used herein refer always to the experimental values.

axis varies, crossing the midplane at the inner and outer ports. At these ports the magnetic flux surfaces are up-down symmetric, whereas for the bottom and top ports they are elongated, lacking any symmetry. The measurements where exclusively done at outer ports with triangularly shaped flux surfaces similar to tokamak geometry.

The magnetic field configuration can be characterised with its local field parameters. As a field line is a three-dimensional curve, at every point in space a tangential vector $\mathbf{t}$ and a curvature vector $\boldsymbol{\kappa}$ can be defined,

$$\mathbf{t} = \mathbf{b} \quad \text{and} \quad \boldsymbol{\kappa} = (\mathbf{b} \cdot \nabla)\mathbf{b} , \tag{5.1}$$

where $\mathbf{b} = \mathbf{B}/B$ is the normalised magnetic field vector. Since the field line lies on the manifold of the flux surface which it spans, also a normal vector $\mathbf{n}$ and a geodesic vector $\mathbf{g} = \mathbf{n} \times \mathbf{b}$ are given at every position on the manifold. The normal vector is, of course, always perpendicular to the surface, pointing outwards in the direction of the minor radius. Therefore, the geodesic vector is also tangential to the surface, pointing into the direction of a curvature of the field line in the surface. For the description of the field line at every point on the manifold, the two curvature components can be used. These are the normal curvature κ_n and geodesic curvature κ_g, given as the projection of the curvature vector $\boldsymbol{\kappa}$ on the respective axis,[4]

$$\kappa_n = \boldsymbol{\kappa} \cdot \mathbf{n} \quad \text{and} \quad \kappa_g = \boldsymbol{\kappa} \cdot \mathbf{g} . \tag{5.2}$$

For a torus geometry the normal curvature is positive on the inboard side and negative on the outboard side. As already discussed in chapter 2.2, negative (bad) curvature is responsible for the destabilisation of micro instabilities and can lead to a wavelike deformation of the isobars, ballooning at the outside (ballooning modes). The geodesic curvature is connected to the parallel return flows since the field lines are bend around the torus and, therefore, link regions of different magnetic field parameters. The charge separation due to the diamagnetic currents is short-circuited along the field lines, leading to the Pfirsch-Schlüter currents, which prevent an electric charging of the plasma. In the case of the zonal flow and the geodesic acoustic mode, where the twisted field lines guarantee a divergence free flow field, the geodesic curvature is responsible for the coupling of the modes and, thereby, their damping (Chap. 3.4.2). But also the radial drift of trapped particles depends on the magnitude of the geodesic curvature. In stellarators this can lead to

---

[4]For geodesic lines, which are the shortest connection on a manifold, the geodesic curvature is zero.

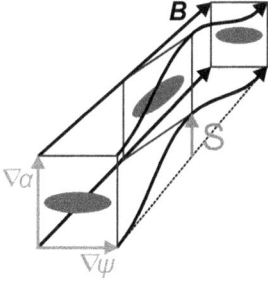

Figure 5.3: Illustration of the effect of magnetic shear on a structure with parallel and radial extent. As the turbulent structures are field aligned, a magnetic shear is transferred to the shape of the structure. [101, 178]

high particle loss rates. Therefore, a minimal geodesic curvature is one of the optimisation criteria for advanced stellarators such as W7-X [174, 175]. For a torus the geodesic curvature has a sinusoidal shape with positive values on the upper half and negative on the lower half, or vice versa. In the TJ-K stellarator this is only found at outer ports whereas for other toroidal angles the trend is more complicated.

The magnetic field is three-dimensional and also the variation in radial direction has to be considered. The rotational transform ι, which describes the twisting of the field lines, depends on the minor radius. Since the turbulent structures are radially extended and mostly field aligned, this represents a background shear which, in general, has a stabilising effect on turbulent modes. For a description of the magnetic shear the magnetic field geometry is considered in a transformed coordinate system $\{\psi, \theta_f, \varphi_f\}$, e.g. Boozer [176] or Hamada coordinates [177], where the field lines are straight lines with the slope of the rotational transform. The toroidal magnetic flux $\psi = \int \mathbf{B} \cdot \nabla \varphi_f \, dV$ denotes the radial coordinate and θ_f, φ_f are the coordinates in the flux surface. A field line can then be specified with the parameter $\alpha := \theta_f - \iota \varphi_f$. Together with the toroidal flux ψ it is connected to the magnetic field vector as $\mathbf{B} = \nabla \psi \times \nabla \alpha$. In this coordinate system the local magnetic shear can be introduced as

$$S := \frac{(\mathbf{B} \times \nabla \psi) \cdot \nabla \times (\mathbf{B} \times \nabla \psi)}{2\pi |\nabla \psi|^4} = -\frac{\mathbf{B} \cdot \nabla \Lambda}{2\pi} , \tag{5.3}$$

where $\Lambda = (\nabla \psi \cdot \nabla \alpha)/|\nabla \psi|^2$ is defined as the integrated local magnetic shear. When there is a local magnetic shear S, radial adjacent field lines are at some point shifted against each other, deviating from their parallel course. This is illustrated in figure 5.3.

5.2 Diagnostics

Probe diagnostics possess very good temporal and spatial resolution but are invasive. They can only be used in plasmas with temperatures of up to a few tens of eV.[5] While there are many different types of probes, Langmuir probes have a comparatively simple layout but can be used to obtain several different parameters like ion saturation current, floating potential, plasma potential, and electron temperature (Sect. 5.2.1).[6] With the combination of multiple probes (Sects. 5.2.2 and 5.2.3), turbulent fluctuations can be resolved spatially and important quantities in turbulence like the Reynolds stress and the vorticity can be obtained.

5.2.1 Langmuir probes

Langmuir probes are made of a tungsten wire (200 µm), which is put through an insulating aluminium oxide ceramic tube. All the probes used in this work are additionally shielded as the tungsten wire runs in a ceramic capillary within a metal tube.[7] Only the tip of the tungsten wire (length 2 mm) is exposed to the plasma and forms the active surface.

When the ungrounded probe is exposed to the plasma, the electrons with their high mobility will charge the probe negatively with respect to the plasma potential ϕ_{pl}. For ambipolarity the currents from both species are in equilibrium and the probe potential is at the floating potential ϕ_{fl}. A bias voltage applied to the probe changes the electron and ion fluxes, which will result in the shown probe characteristic (Fig. 5.4).[8] In the region of strong negative voltage ($U \ll 0$) the current saturates because electrons cannot reach the probe anymore. This results, under the assumption of cold ions ($T_i = 0$), in the ion saturation current

$$I_{i,\mathrm{sat}} = 0.61 e n S_\mathrm{p} \sqrt{\frac{T_e}{m_i}} , \tag{5.4}$$

---

[5] In fusion experiments they are therefore only used in the SOL.

[6] For the measurement of the ion temperature other methods have to be used like tunnel probes, retarding field analyser or spectroscopic methods (Laser induced fluorescence (LIF), charge exchange recombination spectroscopy (CXRS)).

[7] This makes the probe more robust, with less vibrations, and the signals do not exhibit any spikes (random occurring peaks with amplitudes outside the signal span).

[8] The characteristic is normally plotted with reversed y-axis.

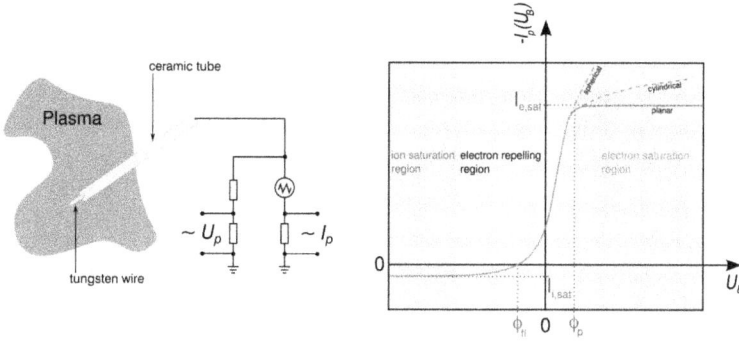

Figure 5.4: Basic principle of a Langmuir probe. With the variation of the biasing voltage the characteristic shown on the right hand side is obtained [166].

which is proportional to the density n but depends also on the probe surface S_p.[9] [10] When the bias voltage is raised, an increasing number of electrons can reach the probe. Only the electrons, Maxwellian distribution supposed, which have enough energy to overcome the potential well can reach the probe, which is expressed by the Boltzmann factor in the electron current

$$I_e = -enS_\mathrm{p}\sqrt{\frac{T_e}{2\pi m_e}} \cdot \exp\left(-\frac{e(\phi_\mathrm{pl} - U)}{T_e}\right).$$ (5.5)

The change in the ion current is small in comparison to the electron current and, therefore, the ion saturation current can be used for the ion current in cold plasmas. The sum of both contributions (5.4) and (5.5) gives the total current to the probe

$$I(U) = I_{i,\mathrm{sat}} + I_e = enS_\mathrm{p}\sqrt{\frac{T_e}{2\pi m_e}}\left\{0.61\sqrt{\frac{2\pi m_e}{m_i}} - \exp\left(-\frac{e(\phi_\mathrm{pl} - U)}{T_e}\right)\right\}.$$ (5.6)

When the probe is on floating potential, both contributions balance and the total current is zero, $I(\phi_\mathrm{fl}) = I_{i,\mathrm{sat}} + I_e = 0$, which results in the relation

---

[9] The effective probe surface depends on the specific parameters of the plasma and, therefore, absolute density values cannot be precisely estimated.

[10] The additional factor is a result of the potential drop in the pre-sheath where the density already decreases to 61 % at the sheath edge [60].

$$\phi_{\mathrm{fl}} = \phi_{\mathrm{pl}} + \frac{T_e}{e} \ln\left(0.61\sqrt{2\pi\frac{m_e}{m_i}}\right) . \tag{5.7}$$

This connects the plasma potential with the floating potential, which can thus be replaced in equation (5.6). Equation (5.6) is valid up to the plasma potential ϕ_{pl} where the electron saturation current $I_{e,\mathrm{sat}}$ is reached. But the current can be increased further since additional electrons from outside the sheath region can reach the probe. In this region the specific shape of the characteristic also depends on the geometry of the probe as depicted in figure 5.4. Only for a planar probe with dimensions large compared to the Debye length $\lambda_{\mathrm{D}} = \sqrt{\epsilon_0 T_e/(e^2 n)}$ the electron current truly saturates.

With an exponential fit to the characteristic the electron temperature can be obtained, whereas temperature fluctuations cannot be resolved with this method.[11] For TJ-K parameters it has been shown that electron temperature fluctuations are small [181]. Therefore, fluctuations in the ion saturation current (measured in the experiment with $-90\,\mathrm{V}$ probe bias) can be associated with density fluctuations ($\tilde{I}_{i,\mathrm{sat}} \propto \tilde{n}$), and floating potential fluctuations are approximately equal to plasma potential fluctuations ($\tilde{\phi}_{\mathrm{fl}} \approx \tilde{\phi}_{\mathrm{pl}}$) [68].

5.2.2 2D-movable probe unit

The 2D-movable probe unit (port O6) enables measurements in the complete poloidal cross section of the plasma with a spatial resolution of $\Delta(R-R_0) = \Delta z = 1\,\mathrm{mm}$. In this work a so-called 3-pin probe is used to obtain radial background profiles of ion saturation current, floating potential, and electron temperature, latter measured with a swept Langmuir probe. Since the effective probe surface S_p (cf. Eq. 5.4) is unknown, the line-averaged density of the microwave interferometer is used to get absolute density values for the ion saturation current profile. The profiles are corrected for a variation of the temperature and the magnetic field, which influences the Larmor radius and, therefore, the effective probe surface, i.e. $S_\mathrm{p} \propto 1/B$. The plasma potential can be calculated according to equation (5.7) or directly determined from the probe characteristic. To this end, the point in the characteristic where the electron saturation regime is reached, i.e. where the curve changes from convex to concave, has to be identified.[12]

With a multi-probe configuration, velocity fluctuations can be measured as the flow velocity is dominated by the $E \times B$-drift velocity (Eq. 2.18). Using

---

[11]Using the conditional sampling technique, the averaged temperature evolution around a trigger time point can be obtained [179, 180].

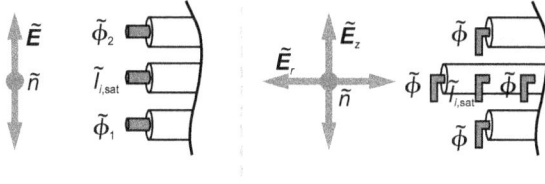

Figure 5.5: Schematic illustration of the 3-pin (left) and the 5-pin probe (right). Because of the configuration of the probes, electric field fluctuations $\tilde{E}$ and density fluctuations $\tilde{n}$ can be obtained at the same time, and, hence, turbulent transport and Reynolds stress can be calculated.

two neighbouring probes (i and $i + 1$) at a distance Δs, the electric field is measured and the perpendicular $E \times B$-drift velocity is given by[13]

$$\tilde{v}^{E \times B} \approx \frac{(\tilde{\phi}_{\mathrm{fl}}^{i+1} - \tilde{\phi}_{\mathrm{fl}}^{i})}{\Delta s\, B} . \tag{5.8}$$

The magnetic field has been considered perpendicular to the electric field. When the middle pin (3-pin probe) is biased to measure ion saturation current, the turbulent cross-field transport can be obtained, at least in the midplane, as product of radial velocity and density fluctuation,

$$\Gamma = \tilde{v}_r \tilde{n} \propto \frac{(\tilde{\phi}_{\mathrm{fl}}^{\theta_{i+1}} - \tilde{\phi}_{\mathrm{fl}}^{\theta_i})}{r \Delta \theta\, B} \tilde{I}_{i,\mathrm{sat}} . \tag{5.9}$$

With another pair of probes (5-pin probe) also the poloidal velocity component is available, which yields the turbulent Reynolds stress[14]

$$R = \tilde{v}_r \tilde{v}_\theta \approx \frac{(\tilde{\phi}_{\mathrm{fl}}^{\theta_{i+1}} - \tilde{\phi}_{\mathrm{fl}}^{\theta_i})(\tilde{\phi}_{\mathrm{fl}}^{r_{i+1}} - \tilde{\phi}_{\mathrm{fl}}^{r_i})}{r \Delta \theta\, \Delta r\, B^2} . \tag{5.10}$$

Due to the 3D structure of the magnetic field, these components are not identical to the normal and perpendicular velocities $\tilde{v}_x$ and $\tilde{v}_y$, with respect to the magnetic flux surface. Compared to the normal-perpendicular

---

[12] Depending on the plasma parameter and the probe geometry this point is difficult measure.

[13] The radial velocity component $\tilde{v}_r$ is directed outwards and the poloidal component $\tilde{v}_\theta$ is chosen to point in the ion diamagnetic drift direction.

[14] In the 2D-plane the measured velocity components are projected on the normal and poloidal unit vector to obtain radial and poloidal velocity, respectively.

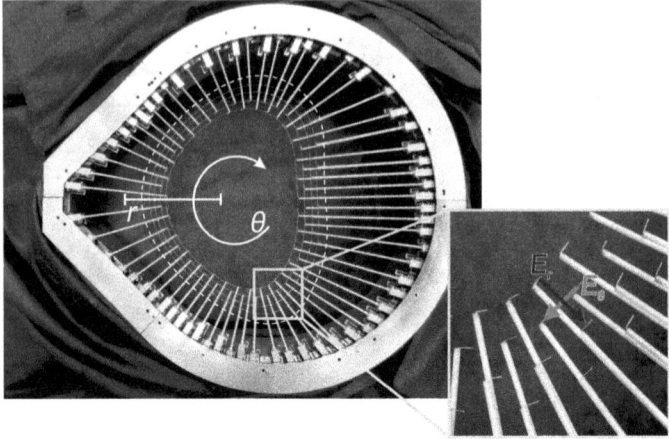

Figure 5.6: Picture of the probe array with Langmuir probes on four flux surfaces. The inlet illustrates the measurement of radial E_r and poloidal E_θ electric field with which the Reynolds stress can be calculated.

Reynolds stress $\tilde{v}_x\tilde{v}_y$, the radial-poloidal Reynolds stress $\tilde{v}_r\tilde{v}_\theta$ is up to approximately 10 % lower on the outboard side.

These movable probe systems only allow point wise measurements where an additional stationary probe has to be used for spatio-temporal analyses. With larger probe arrays the turbulence can be pictured as a whole, resulting in an advanced diagnostic access.

5.2.3 Poloidal Reynolds stress array

Measurements on a complete flux surface are desired since the zonal flow is a mesoscopic turbulent structure, characterised as zonal potential perturbation (cf. Chap. 3.1.1). Therefore, a poloidal probe array was deployed, shown in figure 5.6, which consists of 128 Langmuir probes with 32 probes on each of four neighbouring magnetic flux surfaces (FS 1 to FS 4, counted from inside). In order to get a perpendicular orientation of pairs of probes with respect to the flux surface, the array, as used in [160], was redesigned with the allowance of nonuniform probe spacing. Therefore, the poloidal velocity is measured directly and a phase delay between the velocity components is thus omitted. The array is designed for the outer port O2 ($\varphi = 90°$) with triangular cross section and placed in the confined region just inside the separatrix (dashed white line), where the gradients are steepest. The average poloidal probe spacing is $\Delta x = 1.4$ cm, 1.5 cm, 1.6 cm, and 1.7 cm on the four different flux surfaces at relative radii $R - R_0 = 9.5$ cm, 10.0 cm, 10.5 cm, and 11.0 cm. Also with a spatial uncertainty of 2 mm, the distances are still below the typical structure size of 3 cm to 5 cm [77, 91, 159].

With low capacitance cables the probes are connected to amplifiers with built-in anti-aliasing filters. The data acquisition system allows a sampling rate of up to 1 MHz at a bit depth of 16 bits per sample. Signals of 2^{20} samples (2^{19} samples for 8 GHz discharges) are recorded for all probes simultaneously. For the poloidal probe array it is possible to switch the operation mode for all 128 probes individually from -90 V probe bias to a floating probe measuring ion saturation current or floating potential, respectively. Different biasing schemes can be defined, which in different combinations enable the acquisition of several quantities. The following six schemes were used for the experiments in this work.

1. All probes are set to measure floating potential. This allows the detection of zonal flows and the measurement of velocity fluctuations and related quantities, like Reynolds stress.

2. On FS 1, FS 2, and FS 3 all probes are set to float whereas on FS 4 they alternate with biased probes. The same quantities as in the first mode are obtained, with reduced resolution, but also rough density information can be deduced.

3. Probes on all flux surfaces alternate in their operation mode when going around the circumference. This results in a further reduction of the resolution but simultaneous two-dimensional potential and density measurement in the complete edge region.

4. Sections of the poloidal angle are in the same mode of operation. Gradients of Reynolds stress and density-based pseudo-Reynolds stress [15] can be obtained at the same time.

5. Probes on FS 1 and FS 3 are set to measure floating potential, other probes are biased. This allows the simultaneous measurement of poloidal wavenumber spectra for density and potential.

---

[15] Density fluctuations are treated as potential fluctuations and the pseudo-Reynolds stress is then calculated as in equation (5.10) (see Chap. 8.2.3).

6. All probes measure ion saturation current. Density-based quantities can be calculated with high resolution on all flux surfaces.

The measurement of density and potential fluctuations on a 2D band on the complete circumference gives the unique possibility to directly detect the zonal flow and to measure related quantities. This is for instance the vorticity calculated from the rotation of the velocity field, which is given as

$$\Omega = \nabla_\perp^2 \frac{\tilde{\phi}_{fl}}{B} \ . \tag{5.11}$$

For a zonal flow the poloidally averaged vorticity is unequal zero. Also the flux surface average (indicated by $\langle \cdot \rangle$) of the Reynolds stress, calculated locally as product of the velocity components, can be obtained. Since this is available on different flux surfaces, the radial gradient of the flux surface averaged Reynolds stress can be calculated,

$$\partial_r \mathcal{R} = \partial_r \langle \tilde{v}_r \tilde{v}_\theta \rangle \ , \tag{5.12}$$

which is a measure of the Reynolds stress drive, discussed in section 3.2.

The access to the zonal potential, as the finger print of the zonal flow, is of great value since it directly captures the dynamics of the drift-wave zonal-flow system. In combination with the 2D-movable probe system this allows the visualisation of the turbulent dynamics around the zonal flow occurrence in the complete poloidal cross section (see Sect. 7.2.1).

Chapter 6

Background profiles and turbulence

For turbulence studies it is important to document the underlying equilibrium plasma, which determines the turbulent state. The scaling behaviour of the equilibrium plasma parameters in the experiment TJ-K has been subject of many studies. In this chapter the measurements conducted for this work, using newly designed poloidal limiter (Chap. 5.1.1), are compared to the theoretical results. The first part deals with the spatial profiles of the main plasma parameters (6.1) and how they scale when the experimental parameters are changed (6.2). In the second part (6.3), the scaling of basic turbulence characteristics will be considered. A summary of the main results is given at the end (6.4).

6.1 Background parameter

Two-dimensional profiles of the plasma cross section are demonstrative but, due to the step wise measurement, require long shot durations. Therefore, radial profiles of the main plasma parameters have been recorded for scaling analyses and compared to particle and energy balance studies (Ref. [171]).

6.1.1 Equilibrium profiles

The equilibrium profiles are the result of heating and transport processes. For the plasmas in TJ-K it was shown that the microwave heating leads to centrally peaked density profiles and hollow temperature profiles with the maximum at the separatrix or further out in the scrape-off layer (SOL). Similarly shaped profiles are obtained for plasmas with poloidal limiters but with a narrower radial extent. The 2D profiles of ion saturation current $I_{i,\text{sat}}$ and floating potential ϕ_{fl} for a helium discharge at low magnetic field, where the microwave of 2.45 GHz is resonant, are shown in figure 6.1. All spatial profiles in this chapter are obtained at port O6 with the 2D-movable probe

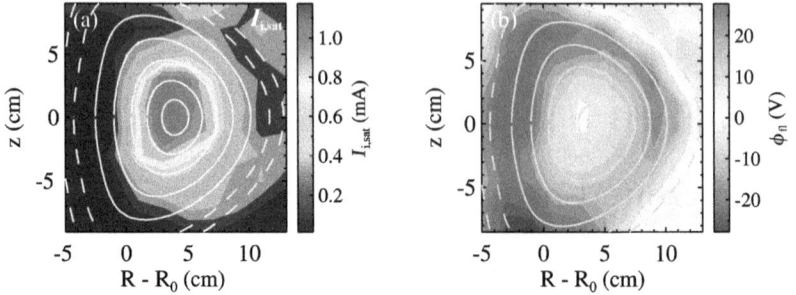

Figure 6.1: Equilibrium profiles of ion saturation current $I_{i,\text{sat}}$ (a) and floating potential ϕ_{fl} (b) in helium at low magnetic field (#9801), measured in the poloidal cross section at port O6. Solid lines show closed flux surfaces whereas dashed lines mark flux surfaces in the SOL.

unit (Chap. 5.2.2). From the ion saturation current profile (a) can be inferred that the density is centrally peaked whereas the floating potential (b) has a pronounced minimum in the edge of the confined region (solid white lines). The floating potential is connected to the electron temperature via equation (5.7). As will be shown further below, the minimum in the floating potential corresponds to a maximum in the temperature located at the same radial position.

In figure 6.2, the radial profiles of density n_e, electron temperature T_e, floating potential ϕ_{fl}, and plasma potential ϕ_{pl} are shown. The profiles are measured from the plasma centre at $R - R_0 = 4.0$ cm to and across the separatrix ($R - R_0 = 12.3$ cm) for all available gases at a neutral gas pressure of $p_0 \approx 8$ mPa and 3 kW microwave heating power (#10259, #10269, #10291, #10309, #10968, #11184).[1] The plotted lines connecting the measurement points are the result of a spline interpolation and given as an illustration of the profile shape. Using the microwave interferometer (port O4), absolute values of the density are obtained. They are corrected for magnetic field and temperature variation (see Chap. 5.2.2) as temperature profiles are available from a swept Langmuir probe, measured simultaneously. The shape of the density (a) and temperature profiles (b) is similar for all gases, and, in principle, the density increases with ion mass. For the scaling of the achievable

---

[1]The pressure refers to the nominal pressure measured by the pressure gauge at the experiment, whereas the real pressure differs by a gas and pressure specific factor. However, the accessible pressure range of each gas does not always overlap.

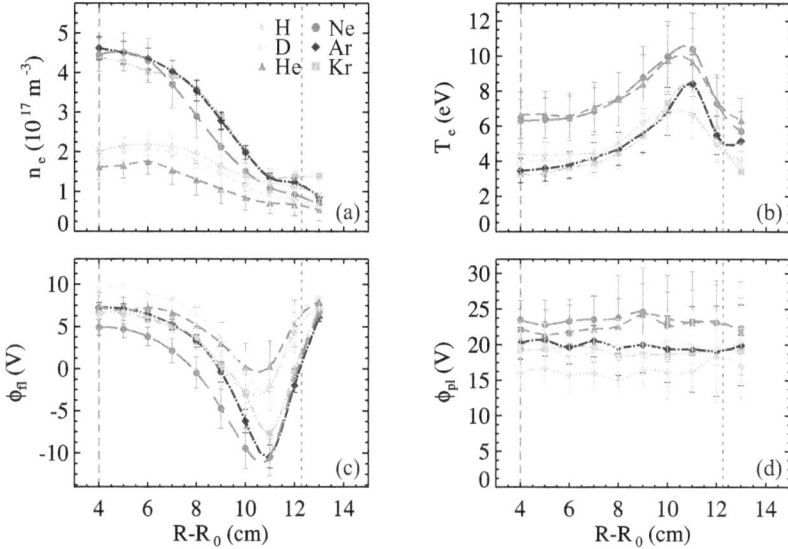

Figure 6.2: Radial profiles of basic plasma parameters for different gases at low magnetic field. The plasma centre is located at $R-R_0 = 4.0$ cm and the separatrix at $R-R_0 = 12.3$ cm, marked by vertical lines. The density n_e (a) is centrally peaked whereas the electron temperature T_e (b) and the floating potential ϕ_{fl} (c) have their extreme values in the edge of the confined region. The plasma potential ϕ_{pl} (d) is relatively flat.

plasma parameters see section 6.1.2. The plasma potential (d), determined as the maximum slope of the probe characteristic, is basically constant with only little variation in the edge region. Together with the electron temperature this explains the shape of the floating potential (c) as it is given as the sum of both quantities.

This demonstrates that the poloidal limiter do not only guarantee a homogeneous SOL but also restrict considerable density and temperature values to the region of closed flux surfaces. Furthermore, the boundary condition for the potential is well-defined as the floating potential should be close to zero at the separatrix.[2]

---

[2] This is an advantage for the external control of the potential profile and, therefore, the radial electric field, which has been demonstrated in biasing experiments [163].

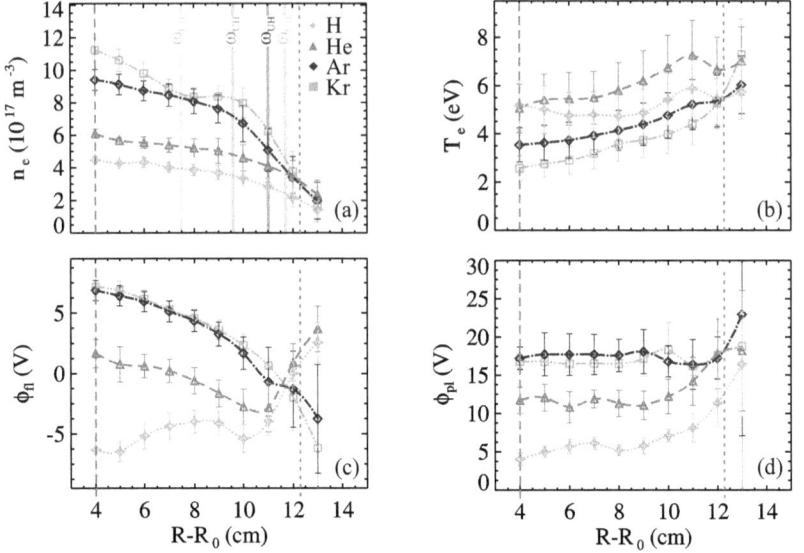

Figure 6.3: Same representation as in Fig. 6.2 but for high magnetic field. Density profiles (a) have similar shape for all gases but profiles of electron temperature (b), floating potential (c), and plasma potential (d) change strongly with the type of gas. The radial position of the upper hybrid resonance ω_{UH} for each gas is marked in figure (a).

It has been found [170, 182, 183] that the plasma is ignited at the electron cyclotron resonance $\omega_{ce} = eB/m_e$ and, when the density is high enough, the microwave is also absorbed at the upper hybrid resonance,

$$\omega_{UH} = \sqrt{\omega_{pe}^2 + \omega_{ce}^2} \,, \tag{6.1}$$

where $\omega_{pe} = \sqrt{e^2 n_e / \epsilon_0 m_e}$ stands for the electron plasma frequency. The resonances lie close to the edge of the confined region or, in the case of the upper hybrid resonance for 2.45 GHz, even in the SOL. Primarily the electrons are heated, hence the electron temperature profile reflects the microwave power deposition profile. With a 1D transport model for electron density and temperature [171], where constant parameters on a flux surface are assumed, the experimentally found shape of the profiles can be reproduced,

$$\frac{\partial n}{\partial t} \approx n_{n0}\, n\, \langle \sigma v \rangle_{\text{ion}} - \nabla \cdot \Gamma_n \,, \tag{6.2}$$

$$\frac{3}{2}\frac{\partial T}{\partial t} \approx \frac{P}{n} - n_{n0} \langle \sigma v \rangle_{\text{ion}} E_{\text{ion}} + \frac{1}{n}\nabla \cdot Q \,. \tag{6.3}$$

The main contribution to the density (Eq. (6.2)) is due to ionisation processes (second term) and particle diffusion (last term). Recombination processes play a minor role and can be neglected. The temperature evolution (Eq. (6.3)) results from the heating term (second term) and is balanced by ionisation (third term) and heat flux (last term). Again, these are the dominant terms where, e.g., neutral gas interaction is not considered. However, with these contributions the shape of the stationary profiles and the scaling of the main plasma parameters, discussed in the next section, can be explained. The heating at the edge of the plasma leads to an inward heat flux which is balanced by ionisation processes. This results in the hollow temperature profile since the energy is gradually lost along the radius. In turn, this energy sink is a source for particles in the plasma centre, subject to diffusion, which explains the centrally peaked density profiles.

In the case of high magnetic field, i.e. 8 GHz microwave heating, the picture is similar (Fig. 6.3). A reduced set of gases is used with similar parameters as above (#10263, #10284, #10297, #10374). The density profile (a) is centrally peaked, and the peak density increases with ion mass. But for argon and krypton the temperature profile (b) has its maximum clearly outside the separatrix. In figure 6.3 (a) the radial position of the upper hybrid resonance ω_{UH} (Eq. (6.1)) for each gas is marked.[3] Due to the high density values, the absorption layer is shifted further outwards. The altered situation is also reflected in the potential profiles (Fig. (c) and (d)). Whereas for hydrogen and helium the minimum in the floating potential is found in the edge of the confined region, for the heavier gases it is clearly outside the separatrix. This substantial change in the equilibrium profiles has to be kept in mind when the scaling of turbulent parameters is considered.

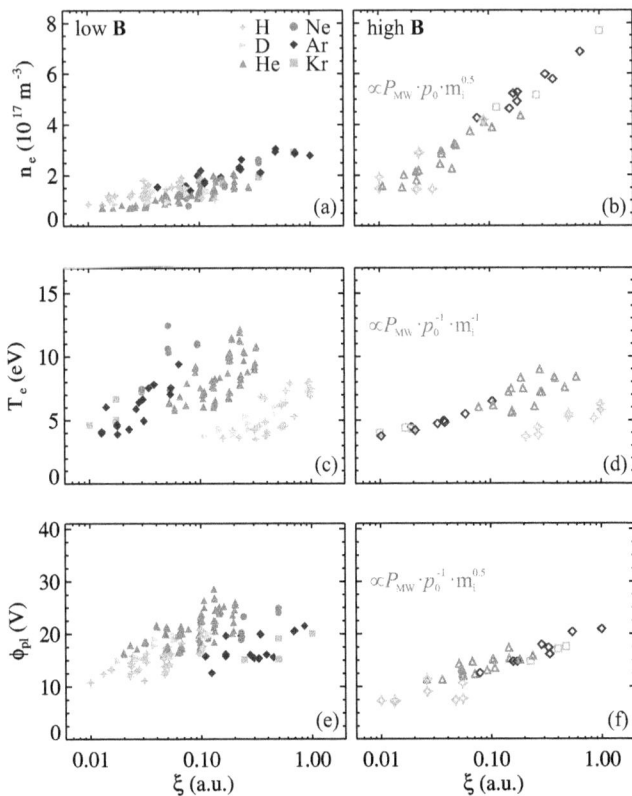

Figure 6.4: Scaling of the main plasma parameters with different experimental settings of injected heating power P_{MW}, neutral gas pressure p_0, and ion mass m_i. The quantities are combined in an experimental control parameter ξ where the proportionality is given in the respective figure. The line-averaged values of density n_e (a,b), electron temperature T_e (c,d), and plasma potential ϕ_{pl} (e,f) are used. Measurements with low magnetic field are shown on the left hand side and high magnetic field discharges are shown on the right hand side, respectively.

6.1.2 Equilibrium trends with experimental control parameters

Because of the flexibility of the experiment, the accessible parameter space is huge. First, the scaling of the main plasma parameters, density, electron temperature, and plasma potential, with more intuitive experimental parameters, like microwave power P_{MW}, neutral gas pressure p_0, and ion mass m_i, is shown. With the concrete dependencies the subsequent scaling of the dimensionless parameters should become more comprehensible.

The transport model, equation (6.2) and (6.3), also allows studies of the parameter dependency. In the simulations [171] the heating power and the neutral gas pressure have been varied for different gases. It was found that an increased heating power results in higher densities due to increased ionisation $\langle \sigma v \rangle_{ion}$, but the temperature should stay constant since the heating term P/n is reduced with density. Because particle diffusion Γ_n and rate coefficient of ionisation $\langle \sigma v \rangle_{ion}$ are gas species dependent, the density increases with ion mass, too. On the other hand, the functional dependency on the neutral gas pressure is more complicated. The density, at least for lighter gases, increases with neutral gas pressure because of the linear dependency of the ionisation term $n_{n0} n \langle \sigma v \rangle_{ion}$ on the neutral gas density n_{n0}. Also the heat sink $-n_{n0} \langle \sigma v \rangle_{ion} E_{ion}$ depends on the density of the neutrals, as a result for higher pressures lower temperatures are obtained. For heavy gases (argon) the ratio of particle and heat transport coefficient depends strongly on the temperature, which results in the moderate variation with neutral gas pressure.

These tendencies can be recovered in the achieved experimental values. The measured values are plotted against the normalised control parameter ξ, which is chosen as the product of microwave heating power P_{MW}, neutral gas pressure p_0, and ion mass m_i with adjusted exponents. In figure 6.4 (a) and (b) the line-averaged density n_e is shown for low and high magnetic fields, respectively. As suggested from the simulations, the density values are scaled with the control parameter proportional to $P_{MW} \cdot p_0 \cdot m_i^{0.5}$. Higher densities are obtained for heavier gases whereas the values for high magnetic field are about a factor of 2 higher. The electron temperature T_e, averaged over the confined region, is scaled with $\xi \propto P_{MW} \cdot p_0^{-1} \cdot m_i^{-1}$ (Fig. 6.4 (c) and (d)). In general, the increase is relatively weak and the temperature values for hydrogen and deuterium do not follow the trend of an increased temperature

---

[3]This is only an indication of the heating position since, in principle, also heating by Bernstein waves has to be considered in this situation [184].

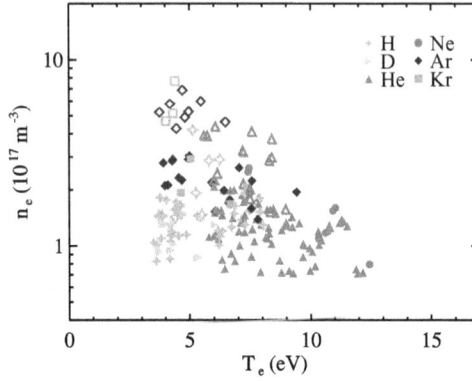

Figure 6.5: Relation between the line-averaged density n_e and the averaged electron temperature T_e. Low magnetic field discharges (filled symbols) and high magnetic field discharges (open symbols) are shown together. The inverse proportionality is visible where temperatures for hydrogen and deuterium are comparatively low.

with lower ion mass. However, this is in line with the simulations which predict temperatures for hydrogen between the temperatures for argon and helium. No big differences are found between low and high magnetic field but, as already visible from the profiles (Fig. 6.2), the values for neon are relatively high, compared to the other gases. Also there are no predictions for the scaling of the plasma potential ϕ_{pl}, the spatially averaged values are shown in the figure (e) and (f) for low and high magnetic field, respectively. A scaling parameter $\xi \propto P_{\mathrm{MW}} \cdot p_0^{-1} \cdot m_i^{0.5}$ has been found to reveal a general trend. Especially for high magnetic field, the averaged plasma potential value increases with ion mass which is the trend expected from figure 6.3 (d). In summary, also with poloidal limiter the experimentally found scaling agrees well with the predictions from the simulation.

For the achievable density an analytic formula can be deduced, which is presented in [185]. The density limit is dependent on the temperature T_e, the heating power P, the neutral gas density n_{n0}, and the ion mass m_i,

$$n_e = \frac{P/(n_{\mathrm{n0}}V_{\mathrm{pl}})}{[(\gamma_D + \frac{3}{2}\alpha_T)T_e + E_{\mathrm{ion}}]\langle\sigma v\rangle_{\mathrm{ion}} + E_{\mathrm{rad}}\langle\sigma v\rangle_{\mathrm{rad}}} \,, \tag{6.4}$$

with the plasma volume V_{pl}, the ratio of thermal to particle diffusivity γ_D, $\alpha_T = T_{e,\text{separatrix}}/T_{e,\text{centre}}$, the ionisation energy E_{ion} with the rate coefficients $\langle \sigma v \rangle_{\text{ion}}$, and the equivalent parameters for the energy loss due to radiation. Because of the many dependencies on experimental parameters, a direct comparison with the theory is complicated. However, a general trend is that higher temperatures go along with lower densities. Figure 6.5 shows the connection of measured densities and corresponding electron temperatures. In principle, the expected trend is visible although the scatter due to the different other parameters is large. Although it has been suggested [157], plasmas with hydrogen do not show highest temperatures.

6.2 Dimensionless parameters

The plasma parameters can be changed over a wide range in the experiment. However, for an investigation of the turbulence more universal parameters are desired, which describe the state of the turbulent system. For this reason dimensionless parameters are introduced in the following which characterise the turbulent length scales and basically reflect the relation between parallel and perpendicular dynamics. Such parameters allow the comparison of the findings with theory and other experiments. In general, the averaged values inside the confined region are plotted with the maximal error shown as error bars.

6.2.1 Mass ratio μ^*

As shown in chapter 2.4, the connection of parallel and perpendicular motion is a key point in plasma turbulence. The timescales of the dynamics perpendicular and parallel to the magnetic field strongly differ since they are covered by the ion dynamics or the electron dynamics, respectively. Therefore, the ratio of these timescales is represented by the electron-ion mass ratio

$$\mu^* = \frac{m_e}{m_i}\left(\frac{R_0 q_{\text{s}}}{L_\perp}\right)^2, \tag{6.5}$$

which is normalised to the ratio of parallel connection length $L_\parallel = R_0 q_{\text{s}}$ to density gradient decay length $L_\perp$. Since the gradient has to be calculated, each density profile is fitted with a function of the type $n_e(r) \propto n_{e,\text{centre}} \cdot \exp\{-(r/\sigma_{\text{pl}})^4\}$, with the peak density $n_{e,\text{centre}}$ and the profile width σ_{pl}. The decay length is then given as $L_\perp = |\nabla_\perp \ln(n_e)|^{-1}$, where the minimal

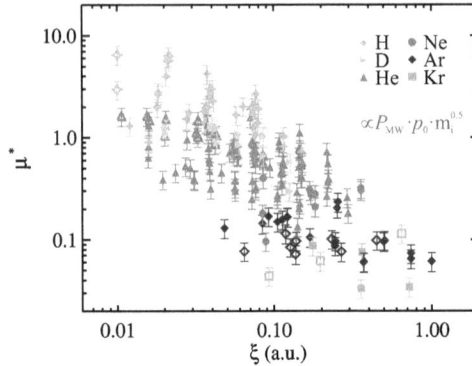

Figure 6.6: Change of the normalised mass ratio μ^* with experimental parameters. The proportionality is given in the figure. All measurements, i.e. low (filled symbols) and high magnetic field (open symbols), are included.

value is used. As the decay length might scale as the density, the same experimental scaling parameter (Chap. 6.1.2, Figs. 6.4 (a) and (b)) is chosen, i.e. $\xi \propto P_{\mathrm{MW}} \cdot p_0 \cdot m_i^{0.5}$. Surely, with ion masses ranging from $m_i^{\mathrm{H}} \approx 1\,\mathrm{u}$ up to $m_i^{\mathrm{Kr}} \approx 84\,\mathrm{u}$, the values of the mass ratio differ almost by two orders of magnitude (Fig. 6.6). Compared to this the dependency on the heating power and neutral gas pressure is relatively weak.

6.2.2 Plasma beta β

The strength of pressure driven interchange instabilities is determined by the plasma pressure (Chap. 2.2.1). Thus, the interchange instabilities should become stronger as the plasma beta $\beta = p/(B^2/(2\mu_0))$, given as ratio of kinetic to magnetic pressure, is increased. Like in the case of the mass ratio, the geometry is accounted for by normalisation through the quotient of parallel connection length $L_{\parallel} = R_0 q_{\mathrm{s}}$ and density decay length $L_{\perp}$,

$$\beta^* = \frac{\beta}{2}\left(\frac{R_0 q_{\mathrm{s}}}{L_{\perp}}\right)^2 \propto \frac{n_e T_e}{(B L_{\perp})^2} \ . \tag{6.6}$$

The scaling of β and the normalised β^* is shown in figure 6.7, where both low and high magnetic field configurations are included. As experimental control parameter, the product $\xi \propto P_{\mathrm{MW}} \cdot p_0^{-1} \cdot m_i^{-1}$ is chosen. The values of

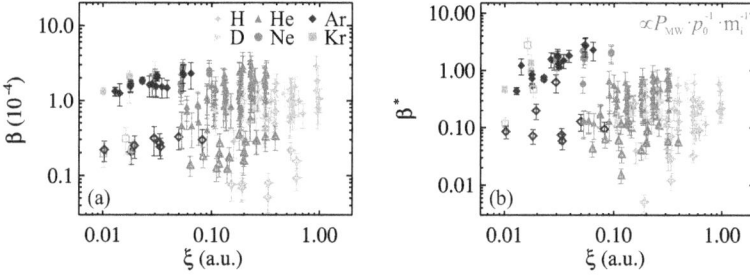

Figure 6.7: Scaling of the plasma beta β (a) and the normalised β^* (b) with experimental parameters. The proportionality is given in the figure. All measurements, i.e. low (filled symbols) and high magnetic field (open symbols), are included.

the plasma-β fall in the same range for all gases as the product of density and temperature remains more or less constant. β can be changed by changing B. In contrast, β^* decrease with ion mass since the decay length $L_\perp$, which enters inversely in (6.6), gets larger for lighter gases. Simulations for TJ-K parameters [88] show that the fluctuations should become more interchange-like when β^* is increased towards the ballooning limit ($\beta^* = 4$), where interchange modes get linearly unstable. However, a pure MHD model does not include Alfvén dynamics, which can counteract this trend (see [88]).

6.2.3 Drift scale ρ_s

The typical size of the turbulent structures becomes larger for heavier gases and is inversely proportional to the magnetic field, like the dependencies of the Larmor radius. This scaling is captured by the drift scale, which is the ion Larmor radius at the electron temperature or rather the ratio of ion sound speed c_s to ion gyro frequency ω_{ci},

$$\rho_s = \frac{c_s}{\omega_{ci}} = \frac{\sqrt{m_i T_e}}{eB} \; . \tag{6.7}$$

This quantity can be related to the minor radius, which is then the dimensionless gyroradius $\rho^* = \rho_s/a$. The drift scale connects the distances in real space to normalised space, which makes it possible to compare turbulent length scales from different experimental conditions. The scalings of ρ_s and ρ^* are shown in figure 6.8 against the control parameter $\xi \propto P_{MW} \cdot p_0^{-1} \cdot m_i^2$.

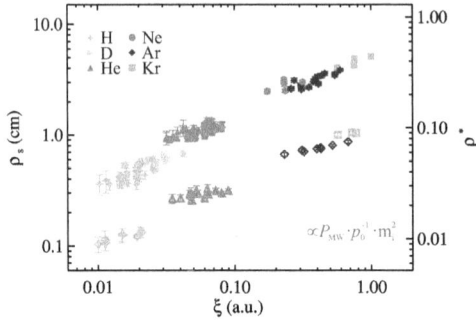

Figure 6.8: Scaling of the drift scale ρ_s and the normalised ρ^* with experimental parameters. The proportionality is given in the figure. All measurements, i.e. low (filled symbols) and high magnetic field (open symbols), are included.

Open symbols show the measurements with high magnetic field, which results in smaller drift scales. The change due to temperature is comparatively small. For the development of homogeneous turbulence, the drift scale should be small compared to the system size. For argon and krypton discharges at low magnetic field this condition is rather poorly satisfied. However, it was shown that the scaling of the poloidal correlation length, which can be regarded as turbulent structure size, scales significantly less than gyro-Bohm like with $\rho_s^{0.43}$ [103, 159, 186]. This issue is addressed again in section 6.3.2.

6.2.4 Collisionality C

As shown in chapter 2.3.4, electron collisions can lead to a phase shift between density and potential fluctuations and thus to a destabilisation of the drift waves. The perturbation of the density potential coupling is described by the collisionality, which is the electron collision frequency ν_e normalised to the electron gyro frequency ω_{ce} compared to the parallel wavelength,

$$C = \frac{\nu_e}{\omega_{ce}} \frac{1}{(k_\parallel \rho_s)^2} \propto \frac{B \, n_e}{k_\parallel^2 \, m_i \, T_e^{5/2}} \; . \tag{6.8}$$

The parallel wavenumber $k_\parallel$ has been obtained from array measurements in a previous work [89, 187] and is normalised with the drift scale ρ_s. Figure 6.9

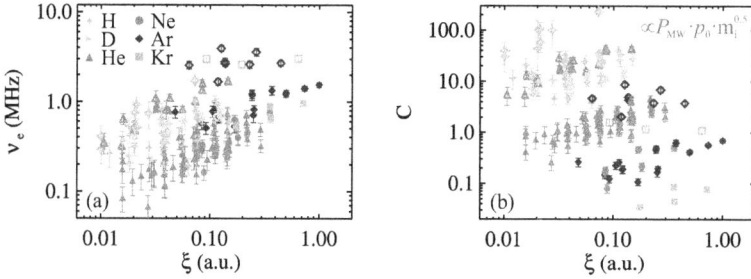

Figure 6.9: Scaling of the electron collision frequency ν_e (a) and the normalised collisionality C (b) with experimental parameters. The proportionality is given in the figure. All measurements, i.e. low (filled symbols) and high magnetic field (open symbols), are included.

shows the scaling of both, collision frequency (a) and corresponding collisionality (b), scaled with the parameter $\xi \propto P_{\mathrm{MW}} \cdot p_0 \cdot m_i^{0.5}$. Since the collision frequency is given as the inverse of the collision time, $\nu_e = \tau_e^{-1} \propto n_e/T_e^{3/2}$, the increase with neutral gas pressure is not surprising. Through the normalised parallel wavelength $(k_\parallel \rho_s)^2$, the collisionality depends on the ion mass which explains the strong decrease for heavier gases. Mainly by changing ion mass, the collisionality can be varied by about four orders of magnitude reaching different regimes of drift wave dynamics. So, especially for hydrogen and deuterium a destabilisation of the drift waves and, therefore, large fluctuation levels are expected (see Sect. 6.3.1).

6.2.5 Parameter dependency

Until now the scaling of the dimensionless parameters was studied with respect to the parameters used in the experiment but in this section the interrelationship between them is examined. This is important for scaling analyses since every parameter can have a specific influence on the turbulent system. In this work the main focus is on scaling with collisionality. Therefore, the relation to the collisionality C is shown in figure 6.10. The mass ratio μ^* (a) increases with collisionality since both depend inversely on the ion mass. But still, the scaling shows the inverse trend for each gas due to the dependency on the density gradient decay length. In general, the normalised beta has a decreasing trend with collisionality. The scatter in β^* for helium and hydrogen is thereby quite large and the values for krypton are a bit off. With

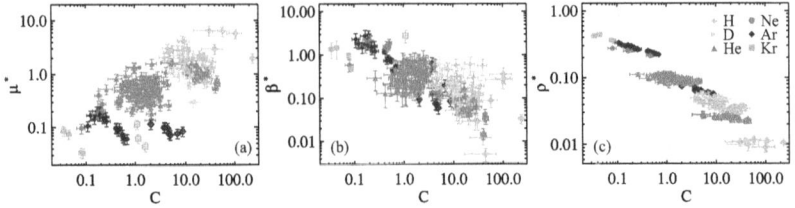

Figure 6.10: Dependency of dimensionless parameters on the collisionality. The mass ratio μ^* (a) increases with collisionality C whereas the plasma beta β^* (b) and drift scale ρ^* decrease. All measurements, i.e. low (filled symbols) and high magnetic field (open symbols), are included.

respect to the drift parameter ρ^* the relationship is relatively clear where a collisional scaling corresponds to an inverse ρ_s-scaling. The comparison of the different dimensionless parameters shows that the discharges are barely dimensionally similar and, therefore, effects ranging from different physical mechanisms can contribute to the scaling of turbulent quantities. This has to be kept in mind for the following scaling analysis.

6.3 Turbulence

Along with the background parameters of the plasma also the turbulence based on it changes. At first (6.3.1) the spatial distribution of the fluctuation level and the scaling with collisionality is investigated. Later on (6.3.2) the spectra of density and potential are analysed in frequency and in wavenumber space.

6.3.1 Fluctuation level

The plasma turbulence suspected in the experiment is driven by gradients in the background plasma parameters and, therefore, the strength of the fluctuations should increase towards the edge. From the time traces recorded with the 2D-unit, the radial dependence of the fluctuation levels of the relevant parameters, density and potential, is obtained. The density fluctuation level is calculated from ion saturation current as $\sigma_n = \tilde{n}_e/n_{e0} = \tilde{I}_{i,\mathrm{sat}}/I_{i,\mathrm{sat}0}$ and the fluctuation level of the potential from the floating potential, i.e. $\sigma_\phi = e\tilde{\phi}/T_e \approx e\tilde{\phi}_\mathrm{fl}/T_e$.

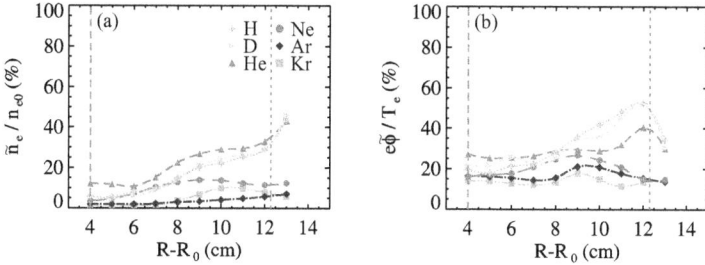

Figure 6.11: Radial profiles of the fluctuation level for different gases at low magnetic field. The plasma centre and the separatrix are marked by vertical lines. In figure (a) the fluctuation level of the density $\tilde{n}_e/n_{e0}$ and in figure (b) of the potential $e\tilde{\phi}/T_e$ is shown. The fluctuation level increases towards the edge and, in general, is higher for smaller ion mass.

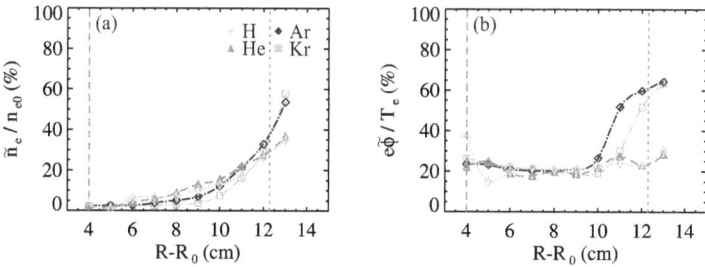

Figure 6.12: Same representation as in Fig. 6.11 but for high magnetic field. Especially in the potential (figure b), argon and krypton show very high fluctuation levels at the edge compared to lighter gases.

Figure 6.11 shows the radial profiles of the fluctuation level for different gases in low magnetic field discharges (same as in chapter 6.1.1). In both cases, density (a) and potential (b), the fluctuation level increases towards the separatrix with a maximum in the edge where the gradients are strongest. This increase with radius is a well known feature of magnetically confined plasmas which can be roughly modelled as $1/n_e$ [188–190]. Also the absolute values of density and potential fluctuation amplitudes are in the same range and increase for lower ion mass. This already suggests a destabilising effect of an increased collisionality, which will be discussed further below.

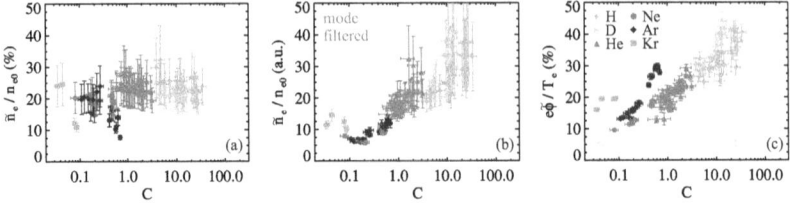

Figure 6.13: Scaling of the fluctuation level with the collisionality for low magnetic field discharges. The fluctuation level for density $\tilde{n}_e/n_{e0}$ (a) and potential $e\tilde{\phi}/T_e$ (c) increase with collisionality C. In the case of the density, this trend becomes clearer (figure b) when the signal is filtered for drift-wave turbulence. See text for further information.

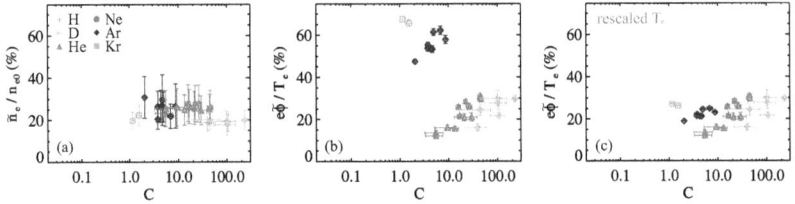

Figure 6.14: Fluctuation level of density (a) and potential (b) for high magnetic field discharges. An increasing trend can be seen but, in the case of argon and krypton, the potential fluctuations are too high. Electron temperatures rescaled by a factor of 2.5 would lead to fluctuation amplitudes similar to the density (c).

The picture is similar for the discharges at high magnetic field, which is shown in figure 6.12. For density (a) and potential (b) the fluctuation level increases towards the edge. But whereas the fluctuation amplitudes for the density only slightly increase with increasing ion mass, the fluctuations of the potential in the plasma edge are much higher for argon and krypton. This seems to be connected to the severe change in the profile form found in section 6.1.1. The standard deviation of the potential in argon and krypton plasmas increases strongly towards the edge but, at the same time, the temperature values are moderate (cf. Fig. 6.3). This suggests that, due to the outward shifted heating position, the temperatures cannot reach high values.

The scaling behaviour of the fluctuation amplitudes with collisionality will now be investigated in more detail. A higher collision frequency can lead to a destabilisation of the drift waves and, therefore, to an increased fluctuation level. For low magnetic field discharges the collisionality scaling is shown in figure 6.13 for density (a) and potential (c). To not only rely on single point measurements, the fluctuation levels are calculated from data obtained with the poloidal probe array which covers the edge region of the confined plasma on the complete poloidal circumference. The fluctuation amplitudes are averaged over the full poloidal angle. The density fluctuation level shows a slight increase with increasing collisionality when different gases are considered. However, the trend for each individual gas is rather reversed. The changing plasma-β, which scales inversely to C, could influence the scaling since it is thought as a drive of interchange modes (cf. Sects. 6.2.2 and 6.2.5). This is supported by the mixing length argument $\tilde{n}_e/n_{e0} = 1/k\,L_\perp$ [191]. Wavenumber k and density decay length $L_\perp$ scale inversely with the collisionality, thus a strong decrease of the fluctuation level with collisionality would be expected when interchange modes would dominate the turbulent fluctuations. With the poloidal probe array it is possible to distinguish different modes in the turbulence by calculating the kf-spectrum (Chap. 4.3.1). Since the drift waves propagate into the electron diamagnetic drift direction (Chap. 2.3.1), they can be identified with modes possessing positive wavenumbers or mode numbers, respectively. When this part of the spectrum ($m_\theta > 0$) is considered for the collisional scaling (Fig. 6.13 (b)), an increase with collisionality is found. The scaling resembles that of the potential (c) closely, clearly suggesting the destabilising effect of electron collisions.

For the high magnetic field discharges (Fig. 6.14) the trend is somewhat similar. The density fluctuations (a) show only a slight increase with colli-sionality with somewhat lower values for hydrogen. A decreasing trend for the individual gas species is missing, pointing to a minor role of interchange turbulence. As in the radial profiles (Fig. 6.12 (b)), very high fluctuation amplitudes in the potential (b) are found for argon and krypton. Due to the shift in the heating position, the electron temperatures inside the confined region seem to be too low to obtain reasonable fluctuation values for the potential (see 6.1.1). But also a poloidal asymmetry in the temperature pro-file exists [166, 167]. Thus, the temperature values from the radial profiles might be underestimated. With the assumption of electron temperatures two and a half times larger for these gases, i.e. $10\,\mathrm{eV} \leq T_e \leq 20\,\mathrm{eV}$, which are still reasonable, fluctuation amplitudes similar to the density would be reached (Fig. 6.14 (c)). Nevertheless, the fluctuation level, with regard to the individual gas species, clearly scales with collisionality suggesting a destabil-

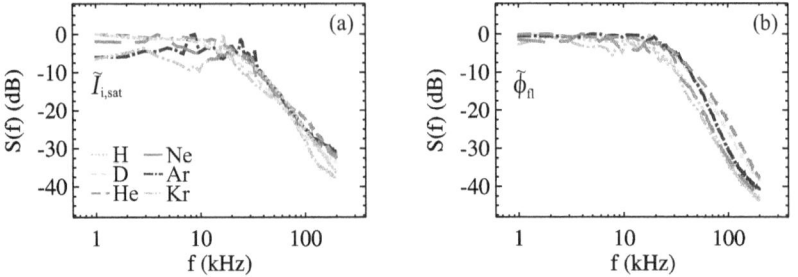

Figure 6.15: Frequency spectra of the ion saturation current $\tilde{I}_{i,\text{sat}}$ (a) and the floating potential $\tilde{\phi}_{\text{fl}}$ (b) for different gases at low magnetic field. The data is measured with the poloidal probe array on the outboard side at port O2. All gases show broad turbulent spectra, however, dominant modes can be found for heavier gases.

isation of the drift waves with increasing collisionality. Overall, the observed scaling is consistent with simulations with the drift-Alfvén turbulence code DALF3 [86] which have been performed for TJ-K parameters [88, 157].

6.3.2 Spectra

A feature of turbulence is the existence of cascades and with it the characteristically shaped turbulent spectra. In figure 6.15 the frequency spectra of density (a) and potential fluctuations (b) are shown for various gases in low magnetic field discharges (#10261, #10268, #10289, #10305, #10969, #11178). Time traces from a probe of the poloidal probe array, located at the outboard midplane, are used. The spectra are not corrected for a potential background velocity and could be modified by an $E \times B$-drift, which, however, should be small. All cases show a broad turbulent spectrum with a flat region up to $10\,\text{kHz}$–$20\,\text{kHz}$, which is followed by an exponential decay up to $200\,\text{kHz}$. The spectra for the light gases are especially smooth while for heavy gases, like argon and krypton, individual modes can be distinguished. This was also observed in [101] where quasi-coherent modes seem to get more intense for higher ion mass and neutral gas pressure. Coherent modes can generally occur in two-dimensional turbulence and are for itself an interesting phenomenon [192–195]. They are like droplets of laminar flow embedded in the background turbulent flow and can alter the turbulent

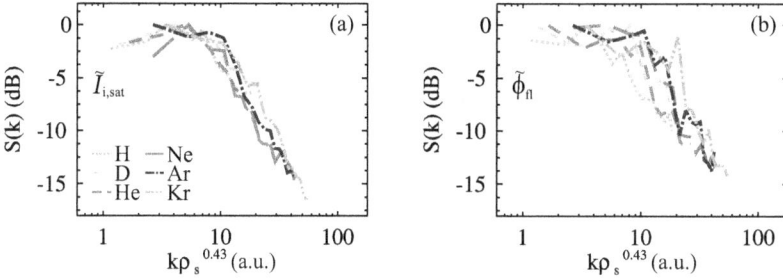

Figure 6.16: Wavenumber spectra of the ion saturation current $\tilde{I}_{i,\text{sat}}$ (a) and the floating potential $\tilde{\phi}_{\text{fl}}$ (b) for different gases at low magnetic field. The wavenumbers are normalised to the structure size which was shown to be proportional to $\rho_s^{0.43}$. For this scaling the density spectra fall together.

spectra significantly. In view of the turbulent zonal flow drive, such modes could capture turbulent energy which is then, depending on their spectral position, not available for the zonal flow. Nevertheless, the measurements in heavier gases will not be excluded since coherent modes are an integral part of the turbulent system.

More relevant than frequency spectra are spectra in wavenumber space, which are directly related to the physical space. In figure 6.16 they are shown for density (a) and potential (b), taken from the same measurements as the frequency spectra. As the frequency spectra, also the wavenumber spectra show a broad inertial range for all types of gas. Since the structure size changes with ion mass, the wavenumbers have to be scaled with the turbulent length scale, typically represented by the drift scale ρ_s (Sect. 6.2.3). In a previous work [103, 159, 186], the structure size in the poloidal cross section has been measured by means of a matrix array with which a detailed scaling behaviour of the density structures could be obtained. It was found that the poloidal correlation length exhibits a scaling between Bohm and gyro-Bohm with $L_\theta \propto \rho_s^{0.43}$. With this scaling factor the density spectra fall together, which confirms the previously found results. In the case of the potential the factor should be a bit smaller for a complete overlap.

Figure 6.17 (a) and (b) show typical wavenumber-frequency spectra (kf-spectra) of ion saturation current and floating potential measured in helium discharges. The broad turbulent spectrum is dominated by contributions at positive wavenumbers indicating a propagation into the electron diamagnetic

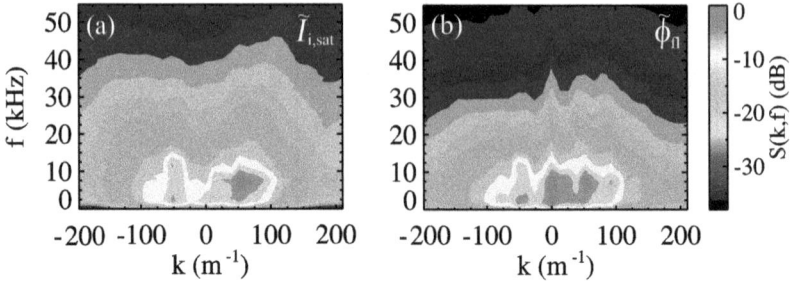

Figure 6.17: Logarithmic wavenumber frequency spectra of ion saturation current $\tilde{I}_{i,\text{sat}}$ (a) and floating potential $\tilde{\phi}_{\text{fl}}$ (b) are shown for low magnetic field discharges in helium (#9984, #9983). Both spectra are dominated by modes with positive wavenumbers, associated with drift waves. The $k_\theta = 0$ mode is the zonal flow, which is not present in the density.

drift direction, which are associated with drift waves. Also modes propagating in the opposite direction, i.e. ion diamagnetic drift direction, can be found in both density and potential. The propagation direction would point to ITG turbulence, but the ion temperatures are suspected to be small. Interestingly, a similar finding was made by L. Cui at CSDX [196, 197] with similar plasma condition as the TJ-K experiment. So far the nature of this mode could not be identified and this topic is still under investigation. A dominant $m = 4$ mode is plausible since drift waves have finite parallel wavelength ($k_\parallel \neq 0$) and the experiment has a rotational transform of $\iota \approx 1/4$ [89, 198]. The $k_\theta = 0$ mode in the potential spectrum is apparent, while not present in the density. This is the signature of the zonal flow, which is known to be a pure potential mode, and it also excludes the possibility of a pure mean background fluctuation since the density is not changed. In comparison with a similar analysis [101] the zonal flow seems more prominent due to the well-defined boundary conditions by the limiter, also in the case of low magnetic field. The characterisation of the zero potential mode will be the topic of the following chapter.

6.4 Summary of the chapter

As the equilibrium profiles are the basis for the development of the plasma turbulence, the analyses in this chapter aim at exploring the accessible parameter space and at verifying basic turbulence characteristics. In contrast to previous works, the measurements have been conducted in plasmas with poloidal limiters while using microwave heating at 2.45 GHz (low magnetic field) and 8 GHz (high magnetic field). Due to the flexibility of the experiment, a variety of gases could be used to extent the parameter range. The main results can be summarised as follows:

- With poloidal limiter the overall plasma parameters and turbulence characteristics are unchanged compared to unlimited discharges, but the profiles are narrower with considerable values of density and temperature restricted to the confined region. The density is centrally peaked and the temperature has a maximum in the edge. In case of high magnetic field argon and krypton show altered profile forms, which seem to be due to the outward shifted heating position as density increases. The profile form and the scaling of averaged density and temperature values fit well to the predictions of a simple transport model.

- The ion mass and the density, which scales mainly with neutral gas pressure, have strongest influence on the dimensionless parameters. The mass ratio μ^*, plasma beta β^*, and drift parameter ρ^* vary in the range of up to two orders of magnitude. In the case of the collisionality C four orders of magnitude are accessible, which makes it possible to gradually change the collisionality in a continuous transition from the adiabatic regime ($C \ll 1$) to the hydrodynamic regime ($C \gg 1$). This is especially important for the investigation of the zonal flow driving mechanism.

- As found in other experiments, the density and potential fluctuation levels increase towards the edge and scale with collisionality. This is expected from simulations for TJ-K parameters and points to a destabilisation of the drift waves through an altered density-potential coupling. A strong influence of a changing plasma beta is not observed. Furthermore, spectra in frequency and wavenumber space show a shape typical for turbulence and confirm previous measurements of the structure size which scales with $\rho_\mathrm{s}^{0.43}$.

- The kf-spectra reveal a dominant zero potential mode, not visible in the density, which is the signature of the zonal flow. In comparison with measurements without poloidal limiter the zonal flow seems to be more prominent.

Chapter 7
Zonal flows

As shown in chapter 6.3.2, the zonal flow, characterised as a zonal potential mode, is an integral part of the plasma turbulence in the experiment. With the special multi-probe configuration deployed in this work the zonal potential is directly accessible and will be studied in more detail in this chapter. First, the basic characteristics of the flux surface averaged potential, as well as their scaling behaviour, will be studied (Sect. 7.1). Including the movable probe unit, the spatio-temporal evolution in the poloidal cross section can be recovered, which is presented in 7.2. Finally, the zonal potential fluctuations are analysed in frequency space where the spectral power distribution is investigated (Sect. 7.3).

7.1 Characteristics of the zonal potential

Poloidal probe arrays are the ideal diagnostic tool to detect zonal flows as such probe configurations directly offer access to the potential fluctuations on the complete poloidal circumference at once. Figure 7.1 (a) shows the poloidally resolved potential fluctuations for 2 ms in a helium plasma at $p_0 = 4.3\,\mathrm{mPa}$ and $P_{\mathrm{MW}} = 2\,\mathrm{kW}$ at low magnetic field. The data has been measured on the flux surface at $R - R_0 = 10.5\,\mathrm{cm}$ (FS 3) with the Reynolds stress array at port O2 (cf. 5.2.3). A poloidal angle of $\theta = 0\cdot\pi$ denotes the outboard midplane, where the outboard side ranges from -0.5π to 0.5π. On the outboard side (low field side) the fluctuations are more intense which correlates with the typical ballooning envelope. The chaotic appearance and the propagation of the turbulent structures into negative θ-direction, i.e. electron diamagnetic drift direction, suggests fully developed drift-wave turbulence (see Sects. 2.3 and 6.3). Below (Fig. 7.1 (b)), the corresponding poloidal wavenumber spectrum, directly calculated from the turbulent fluctuations, is shown. For a better visibility of the $k=0$ mode the mirrored spectrum is plotted as well. At, e.g., $\tau \approx 15.1\,\mathrm{ms}$ a strong zonal mode

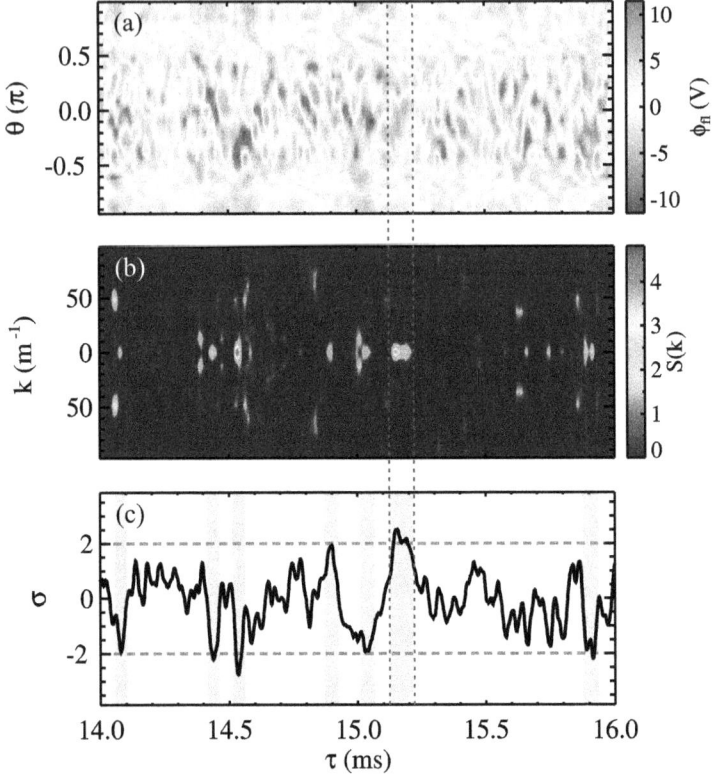

Figure 7.1: Potential measurements in a helium discharge with $p_0 = 4.3\,\mathrm{mPa}$ and $P_{\mathrm{MW}} = 2\,\mathrm{kW}$ at low magnetic field (#10320). The contour plot at the top (a) shows the raw signal of the floating potential fluctuations on a complete poloidal circumference. A poloidal angle of $\theta = 0\cdot\pi$ relates to the midplane on the outboard side. Below (b) is the corresponding wavenumber spectrum. The time trace of the flux surface averaged potential, i.e. $k_\theta = 0$, is shown in (c). Already in the raw time traces the prominent zonal potential mode is visible (e.g. $\tau \approx 15.1\,\mathrm{ms}$).

emerges which lasts up to 100 µs. Also in the raw signal (Fig. 7.1 (a)) the zonal potential mode is indicated, which underlines the significance of the zonal mode in the turbulent system. The occurrence of the zonal potential mode seems to involve modes with higher wavenumbers. They get strong just before the zonal mode arises which is the typical dynamic of the drift-wave zonal-flow system (see 3.3 and [38, 101]). In figure 7.1 (c) the signal of the flux surface averaged potential, equivalent to the time trace of the $k=0$ mode, is shown. For the zonal flow occurrence at $\tau \approx 15.1$ ms the flux surface averaged potential is positive and gets larger than 2σ. But also events where the zonal potential is strongly negative occur, which correspond to zonal flows with reversed flow direction. The flux surface averaged potential is directly the signal of the zonal flow, but, as all turbulent signals, it strongly fluctuates around its mean since it is also subject to random fluctuations and noise. Often large fluctuations in the zonal potential are identified as zonal flows, however, an exact discrimination is difficult.

In the next step, the flux surface averaged potential signal will be examined further as it contains all the dynamics of the zonal flow. Figure 7.2 (a) shows the PDF of the zonal potential. The signal distribution is close to a Gaussian distribution, which is shown in the figure as dashed curve for comparison. This demonstrates the turbulent nature of the zonal potential with the actual zonal flow embedded into the turbulent dynamics. Large fluctuations do not dominate the signal as the kurtosis $\mathcal{K}$ is close to zero. The vanishing skewness $\mathcal{S}$ shows the symmetry of the distribution, which implies that neither positive nor negative zonal flows are preferred. Here, a positive potential fluctuation is denoted as positive zonal flow and vice versa. For the detection of single zonal flow events the time trace is scanned for time points where a trigger condition is met (Chap. 4.6). The distribution of the number of trigger events depending on the trigger value is shown in figure 7.2 (b). Throughout the work, the trigger value is specified in terms of the standard deviation σ. A negative standard deviation signifies triggering on negative amplitudes, i.e. falling below the trigger value. As trigger window, the time which is excluded after a trigger time point, 128 µs are used. The figure shows that, also when only trigger events alone are counted, the distribution is close to a Gaussian. With more than 10^3 events at a trigger value of 2σ the number of realisations is high enough to cover the statistical average.

For the trigger value $|2\sigma|$, which is normally used, the waiting time statistics, together with the cumulative probability, is shown in figure 7.2 (c). The abscissa is scaled to the trigger window (here 128 µs). A maximum is found for one and a half of the window time length which is roughly 200 µs,

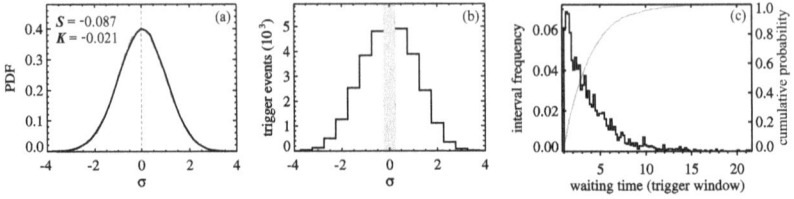

Figure 7.2: Statistical distribution of the flux surface averaged potential signal (#10320). Figure (a) shows the power distribution function (solid line) together with the normal distribution (dashed line). The distribution of the trigger events depending on the trigger level is shown in (b). All distributions are close to a Gaussian distribution. In figure (c) the corresponding waiting time statistics for a trigger value of $|2\sigma|$ is plotted.

then the frequency of occurrence decreases exponentially for longer intervals. The most frequently occurring waiting time corresponds to a fast repetitive occurrence of the zonal flow which can also be observed in figure 7.1. The characteristics are close to ideal fluctuations, and it is clear that the zonal flow, or a specific pattern, is hard to distinguish from noise or a random fluctuation which is not driven by the ambient turbulence. Nevertheless, the flux surface averaged potential is the best diagnostic access to the zonal flow, and reaching a threshold value is used as trigger condition in the rest of this work. It would be interesting to apply more sophisticated methods like clustering algorithms with which different dynamics of the turbulent system could be distinguished but this is left for future work (Chap. 10).

For a trigger value of 1σ the collisional scaling of the number of trigger events (a,c) and the corresponding mean waiting time (b,d) are shown in figure 7.3. At the top are the pictures for low magnetic field and below for high magnetic field. The trigger counts are comparable as the measurement time is twice as long for low magnetic field. In both cases the count of trigger events decreases for lower collisionality C, which is more pronounced for high magnetic field. As a decreased number of counts results in a longer time between them, the waiting time shows the inverse trend and increases for lower collisionality. Since the trigger value is always specified with respect to the standard deviation, this shows a change in the dynamic of the system. In general, the turbulent dynamics seem to get slower for lower collisionality and higher ion mass.

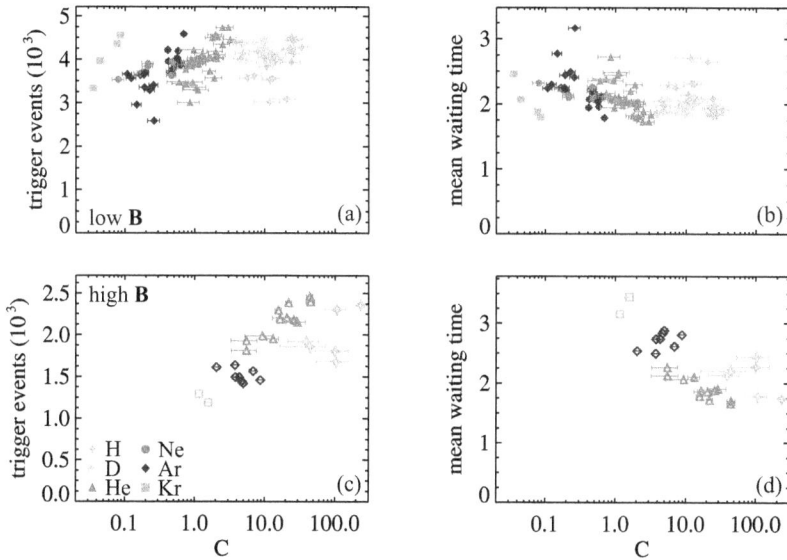

Figure 7.3: Collisional scaling of trigger events for low (a,b) and high magnetic field (c,d). The total number of trigger events (a,c), where all occurrences above 1σ are counted, seems to decrease with collisionality C. Therefore, the interval time between the events increases accordingly.

7.2 Structure and dynamics

With the conditional averaging and cross-correlation methods the spatio-temporal evolution of turbulent structures can be obtained. For the analysis of the zonal flow, the flux surface averaged signal is used as trigger or reference signal. At first (7.2.1), the dynamics in the poloidal cross section is presented, which is obtained by the combination of the movable probe unit and the poloidal probe array. The temporal evolution on the complete flux surface is shown subsequently (7.2.2).

7.2.1 Dynamics in the poloidal cross section

The poloidal probe array (Chap. 5.2.3) is situated in the edge region of the confined plasma on the complete circumference in the poloidal cross section. This results in a good spatial resolution in the poloidal direction but the resolution is rather poor in the radial direction. On the other hand, the 2D-movable probe unit (Chap. 5.2.2) has a very good spatial resolution but needs a reference probe for spatio-temporal analyses (cf. Chap. 4.6). Both measurement techniques were combined to reliably detect the zonal potential perturbation and to obtain the turbulent fluctuations in the poloidal cross section. For this purpose the data of the array is recorded simultaneously for each step of the movable probe. With the large number of probes, this is a challenging task for the data acquisition system, which limits the number of steps and, therefore, the achievable spatial resolution.

For a half profile ($R - R_0 \in [5\,\mathrm{cm}, 12.5\,\mathrm{cm}]$) at port O6 the potential and density fluctuations are shown for three time points relative to the trigger position in figure 7.4. The potential is shown at the top (a–c) and the density, for the same time points, below (d–f). The data of the movable Langmuir probe is conditionally averaged with 2σ in the zonal potential fluctuations from the array. Subwindows are centred on the maximum near the trigger event. At the first time point at $\tau = -94\,\mu\mathrm{s}$, well before the trigger time point ($\tau = 0\,\mu\mathrm{s}$), a prominent turbulent structure is visible in figures (a) and (d). It appears in the potential and the density with a small cross-phase and propagates into the electron diamagnetic drift direction. It can thus be identified as a drift wave (Sect. 2.3). In the next frame ($\tau = 7\,\mu\mathrm{s}$), shortly after the trigger condition is reached, in the density (e), again, a localised structure is visible whereas in the potential (b) the appearance of a zonal mode can be seen in the edge of the confined region. This is the expected structure of a zonal flow which exists in the potential but not in

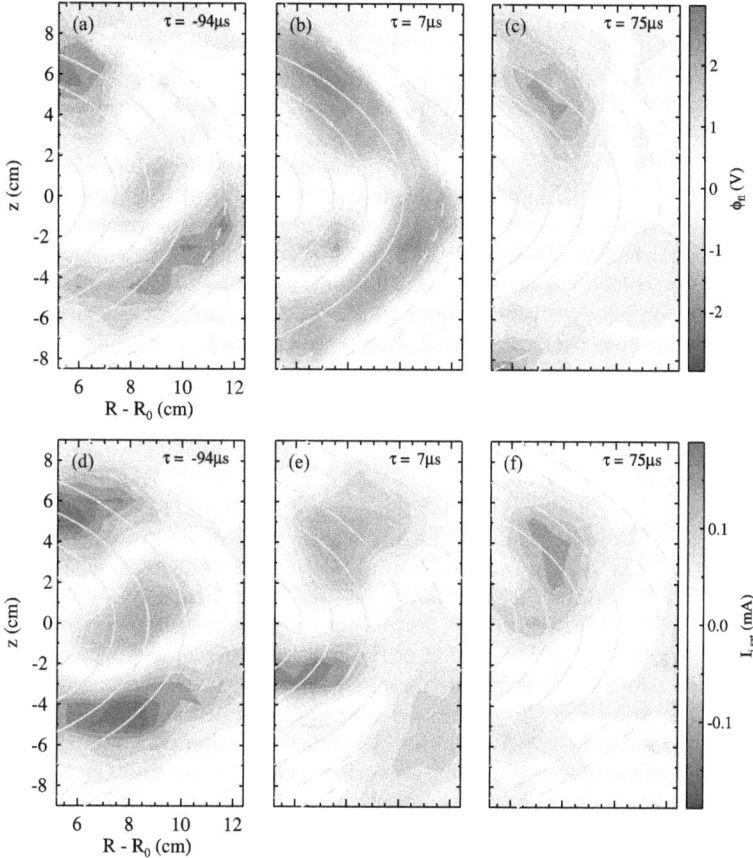

Figure 7.4: Temporal evolution of the potential (top row) and the density (bottom row) in the poloidal cross section. The figures are ordered chronologically from left to right with the time relative to the trigger time. The signals from the 2D-movable Langmuir probe are conditionally averaged with ϕ trigger from the surface averaged potential measured with the poloidal probe array at the distant port O2. Flux surfaces in the SOL are plotted as dashed lines.

the density [33]. With a radial extent of roughly $2\,\mathrm{cm}$ ($k_r = 50\,\mathrm{m}^{-1}$) the structure size is in the range of the dominant drift wave structures. Soon after the trigger condition is reached the zonal flow has vanished and another drift wave structure still remains. In the scrape-off layer (SOL), marked as dashed lines, turbulent structures propagating into ion diamagnetic drift direction (upwards) are found. These so-called blobs have been characterised in various studies [166, 180, 199, 200] but it is unclear if they are linked with the zonal flow occurrence. As this technique combines measurements at two different toroidal positions with a separation of $\varphi = 120°$, this analysis shows the existence of long-range correlations also in the toroidal direction. Thus, the zonal potential measured with the array seems indeed to correspond to a potential perturbation on a complete flux surface ($k_\theta = k_\varphi = 0$) with a finite radial extent ($k_r \neq 0$) typical for zonal flows. After the spatial structure of the zonal flow has been analysed the temporal evolution is examined in more detail in the next section.

7.2.2 Dynamics on a flux surface

To obtain the averaged evolution around the zonal flow occurrence the signal of the poloidal probe array is conditionally averaged with ϕ trigger from the surface averaged potential, as done in the analysis in 7.2.1. Since zonal flows exist as positive and negative potential fluctuation, both cases, calculated with oppositely signed trigger values, are shown in figure 7.5. As before, the individual subwindows are centred on their respective maximum. At the top is the temporal evolution of the flux surface averaged signal $\langle \phi_{\mathrm{fl}} \rangle_{\mathrm{fs}}$ and below are the potential fluctuations poloidally resolved. In contrast to figures 7.1, figure 7.5 shows the averaged evolution of the potential fluctuations. This is not the dynamic of a zonal flow in general but more the evolution of the dominant zonal perturbation, where the individual zonal flow event can significantly differ from the average. This can be seen with the conditional deviation σ_{CA} [201], as a measure of the goodness of the average, which is defined as the standard deviation σ of the difference of the averaged signal $\langle X \rangle_{\mathrm{CA}}(\tau)$ and the single realisations $[x(t_i + \tau)]_{i=1\ldots N}$,

$$\sigma_{\mathrm{CA}}(\tau) = \frac{\sigma\big([\langle X \rangle_{\mathrm{CA}}(\tau) - x(t_i + \tau)]_{i=1\ldots N}\big)}{\sigma(X)} \,. \tag{7.1}$$

The values of the conditional deviation, without normalisation, are shown in the figures (a,b) as shaded area. Around the trigger time point the values are small whereas they are in the range of the overall standard deviation of

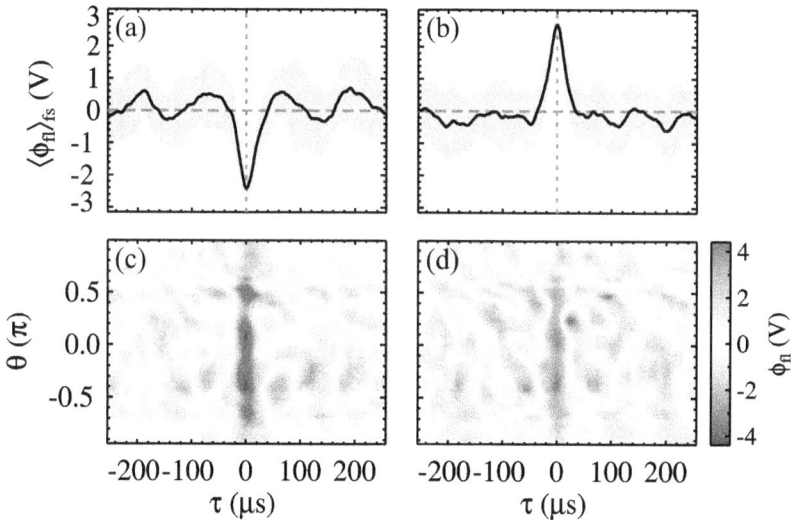

Figure 7.5: Conditionally averaged time evolution of the potential around the trigger time point ($\tau = 0\,\mu s$). Figures at the top (a, b) show the flux surface average signal and below (c, d) is the evolution on the complete flux surface. For the pictures on the left side the signal is triggered on positive events whereas on the right it is triggered on negative ones.

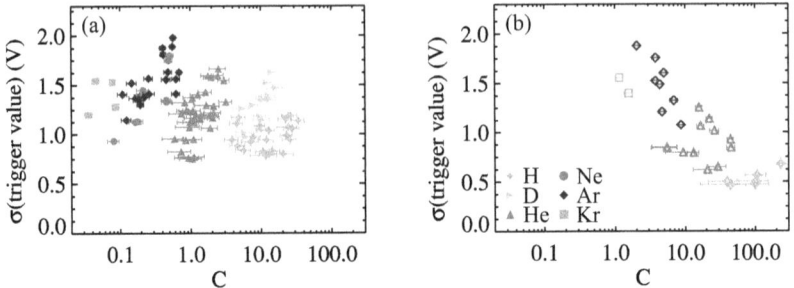

Figure 7.6: Collisionality scaling of the standard deviation σ of the distribution of the zonal flow amplitude. On the left hand side (a) is the scaling shown for low magnetic field and on the right hand side (b) for high magnetic field. In general, the variance is increasing for decreasing collisionality.

the signal in the rest of the time. This shows that the single zonal flow event must have a quite different dynamic, however, the conditional average still exhibits features characteristic to all realisations. In the contour plots (c,d) the propagation of the turbulent structures in the negative θ-direction is clearly visible. This correlation demonstrates the importance of these structures, thought of as drift waves, for the zonal flow evolution. At $\tau = 0\,\mu s$ the potential on the complete flux surface is positive, or negative, respectively, which corresponds to a poloidal wavenumber of $k_\theta = 0$. The (averaged) life time of the zonal flow, in both cases, is relatively short with $\Delta\tau \approx 50\,\mu s$.

Key parameters like the zonal flow amplitude and the life time can now be extracted from the flux surface averaged potential. The zonal flow amplitude, identified as the peak value of the single event, has a the trigger value as minimum but follows, of course, a distribution. Figure 7.6 shows the collisionality scaling of the standard deviation of the zonal flow amplitude. For low (a) and high (b) magnetic field the distribution gets broader for lower collisionality, which is especially clear in the second case, implying the increased occurrence of strong zonal flows. This already indicates that the zonal flow power increases with lower collisionality, meaning a more adiabatic electron response. This will be subject of the discussion in section 7.3.2. Noticeable are the relatively high values for deuterium in comparison with the values for hydrogen. Although the discharges have comparable collisionalities, the zonal flow power is, in nearly all cases, higher in deuterium. This

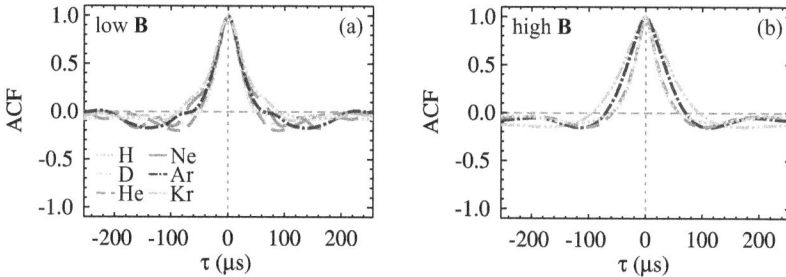

Figure 7.7: Auto-correlation function (ACF) of the flux surface averaged potential for different gases at low (a) and high (b) magnetic field. The ACF becomes broader with increasing ion mass.

phenomenon could be related to the isotope effect, which has been observed at other fusion experiments, meaning better confinement in D than in H.

Next, the scaling of the zonal flow life time is examined. To extract an averaged value for each measurement the auto-correlation function (ACF) of the flux surface averaged potential is calculated. Figure 7.7 shows the auto-correlation function for all gases used at low (a) (#10377, #11195, #10306, #10295, #10277, #10262) and high (b) (#10264, #10284, #10301, #10372) magnetic field. A variation of the peak width with the gas species is visible, which becomes wider for heavier ion mass.

For a quantitative analysis the auto-correlation time (ACT), i.e. the time until the signal has dropped below the value $1/e$, is calculated for each measurement. The actual life time of the zonal flow would then be at least twice the auto-correlation time. The collisionality scaling of the auto-correlation time is plotted in figure 7.8 for low (a) and high (b) magnetic field. Although the variation is close to the temporal resolution of the measurement, an increase of the correlation time with lower collisionality is found in both cases. For low magnetic field this increase is mainly restricted to the scaling within a gas species. In contrast to the zonal flow amplitude (Fig. 7.6 (a)), the deuterium measurements do not show higher values for the zonal flow life time but are in the same range as the values for hydrogen. In summary, a lower collisionality seems not only to lead to a higher amplitude but also to a longer life time of the zonal flow (even though the trend, in both cases, is relatively weak for the overall scaling at low magnetic field).

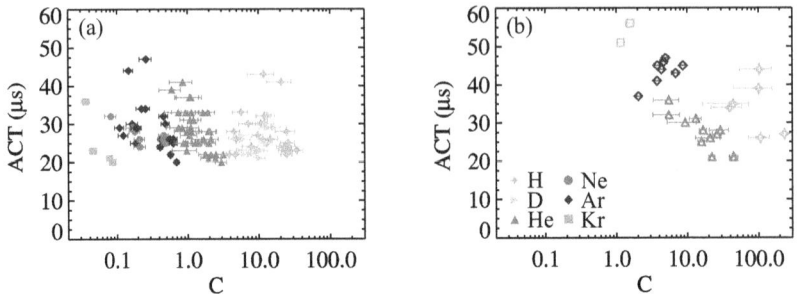

Figure 7.8: Scaling of the auto-correlation time (ACT) for low (a) and high (b) magnetic field with collisionality C. Especially for discharges with high magnetic field the auto-correlation time increases strongly for lower collisionality C.

7.3 Spectral analysis

The zonal potential, available at high temporal resolution, is now analysed for its spectral distribution. This is part of the first section 7.3.1. In the second part (7.3.2) the collisional scaling of the zonal flow contribution to the spectrum, relative zonal flow power, is considered, which has been published in [202].

7.3.1 Frequency distribution

Previous investigations [101, 167] showed that the zonal potential fluctuations have a frequency below 10 kHz. This is in accordance with the basic requirements for zonal flows (cf. Sect. 3.1.1) where they are assumed to be clearly separated from the ambient turbulence. In figure 7.9, the spectral distribution of the zonal potential fluctuations are shown for various gases at low (a) and high magnetic field (b). The frequency spectra are calculated from sub-series with a window length of roughly 1 ms which are averaged 1024 times or 512 times for low and high magnetic field, respectively. In both cases the main part of the spectral power is concentrated in the low frequency range below 8 kHz. For high magnetic field the spectra are smooth and the absolute power of the fluctuations increases with ion mass. The trend is similar, however, not as clear for low magnetic field. In this configuration a second maximum is visible at higher frequencies ($\approx$ 18 kHz) for the heavy gases neon, argon, and krypton. This is a (relative) fast fluctuation of the

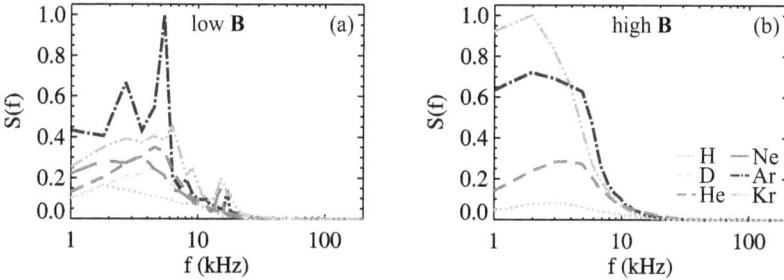

Figure 7.9: Frequency spectrum of the flux surface average potential. The spectra for different gases for low (a) and high (b) magnetic field are shown. The main spectral power is concentrated below 8 kHz, which is the expected frequency range of the zonal flow.

zonal potential which might be connected to the GAM oscillation (see 3.1.2). However, their contribution would be rather small. This discussion is continued in chapter 9 where the different modes, interacting with the zonal potential, are analysed.

Simulations with collisional trapped electron mode turbulence [203] suggest that the frequency spectrum of the zonal potential should become wider for a nonadiabatic electron response, i.e. high collisionality or here smaller ion mass. This is however difficult to estimate as the variation of the spectral width is small. Furthermore, no clear trend seems visible for the spectral position of the maximum. Since a gas specific dependency seems to exist, the collisional scaling is analysed additionally. For both configurations, the maximal spectral power against the collisionality is shown in figure 7.10. The trend is similar to the one shown in figure 7.6 due to the fact that both are related quantities. Again, the increasing trend with decreased collisionality is especially clear for high magnetic field and the values for deuterium, where the spectral zonal flow power is relatively high, stick out.

7.3.2 Collisionality scaling of spectral power

As already visible from the results so far, the collisionality has an influence on the drift-wave zonal-flow system. Of special interest is the scaling of the zonal flow power. An increased collisionality cumbers the electron response (Sect. 6.2.4) which influences the driving mechanism of the zonal flow

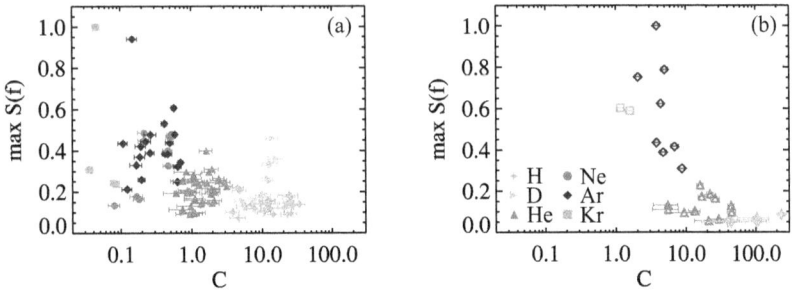

Figure 7.10: Collisionality scaling of the maximal contribution to the flux surface averaged potential spectrum. On the left hand side is the scaling with discharges at low magnetic field (a) and on the right for high magnetic field (b). In both cases the spectral power increases with lower collisionality C.

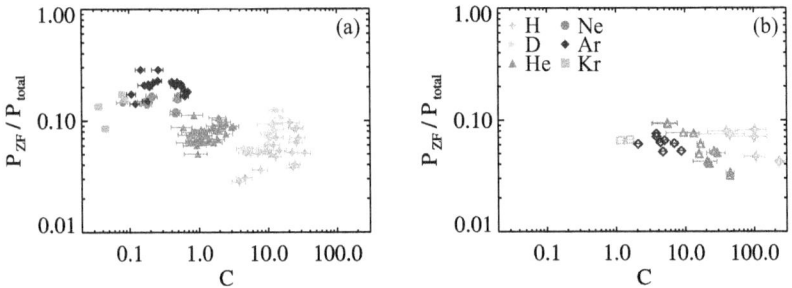

Figure 7.11: The scaling of the relative zonal flow power with collisionality. For a clearer representation, the scaling for low magnetic field is shown on the left hand side (a) and for high magnetic field on the right hand side (b). The spectral contribution of the zonal flow increases with lower collisionality.

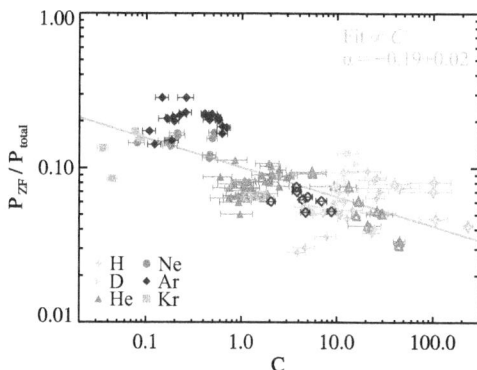

Figure 7.12: Same representation as in Fig. 7.11 with a power law fit to all data points. The fit parameters are given in the figure.

(Sect. 3.2). Figures 7.6 and 7.8 show that the absolute amplitude and the life time of the zonal flow increase with lower collisionality. Especially for the low magnetic field configuration the trend is not really conclusive. However, the accessible collisionality range covers four orders of magnitude and the turbulent state changes strongly. As the overall turbulence depends on the background gradients, not the absolute zonal flow power is interesting, but the relative contribution to the turbulent spectrum. Therefore, the relative zonal flow power is calculated as

$$P_{\mathrm{ZF}}/P_{\mathrm{total}} \;=\; \sum_{f \le 8\,\mathrm{kHz}} S_\phi(k=0,f) \,/\, \sum_{k,f} S_\phi(k,f) \qquad (7.2)$$

from the wavenumber frequency spectrum $S_\phi(k,f)$ of the potential (cf. Chap. 4.3.1). As indicated above, for the zonal flow component only the low frequency bandpass filtered spectral power is used. For a better visualisation figure 7.11 shows the collisionality scaling of the relative zonal flow power for low (a) and high (b) magnetic field separately. In the adiabatic limit ($C \to 0$), the zonal flow contribution to the complete spectrum strongly increases, and the zonal flow power reaches values of up to 29 % of the total turbulent spectral power. For high magnetic field this trend is especially clear when the individual gas is considered. The large scatter in the case of low magnetic field might be connected to the change in other relevant parameters, like plasma-β (Sect. 6.2.5). Nevertheless, the overall trend of

129

the relative zonal flow power cannot be explained by an increased plasma-β as this should have either no impact on the zonal flow or should decrease the relative zonal flow power as the total power enters inversely in equation (7.2). To deduce a quantitative assertion, the collisionality scaling of the relative zonal flow power is fitted with a power law $\propto C^\alpha$, where a value of $\alpha = -0.19 \pm 0.02$ is found when all measurements are considered (see Fig. 7.12). For low magnetic field measurements alone the fit yields a slightly higher value of $\alpha = -0.23 \pm 0.02$.

7.4 Summary of the chapter

The poloidal probe array enables to directly access the flux surface averaged potential, which is though of as the zonal flow mode. Together with the 2D-movable probe unit, using the conditional averaging technique, the zonal flow evolution was visualised in the poloidal cross section. Key parameters were extracted from the zonal potential and scaled with the collisionality. The analysis of the zonal potential could reveal the following points:

- The distribution of the flux surface averaged potential, as well as the number of zonal flow events, is close to a Gaussian. And the corresponding waiting time decreases exponentially with a maximum at $200\,\mu\text{s}$ when triggered on $|2\sigma|$. This already illustrates the dynamic behaviour of the drift-wave zonal-flow system where the zonal flow appears in a burst like manner.

- Positive and negative potential fluctuations, meaning opposite flow directions, occur equally distributed. The averaged temporal evolution of both flow patterns are very similar and correspond to a single burst, where the potential is positive or negative on the whole flux surface. In the poloidal cross section the zonal flow appears as a ring like structure with a radial extent k_r in the range of the dominant drift waves. The measurements at two distant toroidal positions clearly demonstrate the 3D structure of the zonal flow.

- The frequency spectra of the zonal potential show that the major power of the zonal flow is concentrated at frequencies below $8\,\text{kHz}$. A spectral broadening with higher collisionality or a shift of the maximal frequency is not observed but a collisionality dependence is indicated.

- The scaling of different zonal flow parameters with collisionality revealed a consistent trend. With decreased collisionality the zonal flow amplitude as well as the zonal flow life time increase. The relationship gets clearer when the relative zonal flow contribution, defined as zonal flow power compared to the complete turbulent power, is considered. Values of up to 29 % are reached and a power law fit (C^α) yields $\alpha = -0.19 \pm 0.02$.

Chapter 8

Reynolds stress drive

The Reynolds stress is a key quantity in turbulence research as it captures the interaction between the turbulent structures. In the context of the zonal flow, a radial gradient of the flux surface averaged Reynolds stress $\mathcal{R} = \langle \tilde{v}_r \tilde{v}_\theta \rangle$, as indicated by the brackets, drives the poloidal flow (cf. Chap. 3.2). However, this does not imply that the Reynolds stress and the resulting local gradients are homogeneously distributed on a flux surface, especially since the underlying plasma turbulence depends strongly on the poloidal angle. With the Langmuir-probe array, the Reynolds stress can be measured poloidally resolved and a direct estimate of the local and flux surface averaged radial Reynolds stress gradient is possible. Chapter 8.1 is dedicated to the spatial structure of the temporal mean Reynolds stress, here referred to as background Reynolds stress, where the dependence on the magnetic field geometry (Sect. 8.1.1) and the connection to the velocity distribution (Sect. 8.1.2) is studied. Subsequently (Chap. 8.2), the dynamic of the Reynolds stress and, especially, the connection to the zonal flow (Sect. 8.2.2) is shown. Results of this chapter have been published in [204].

8.1 Background Reynolds stress

From the poloidal momentum balance equation (3.17) follows that a radial gradient in the Reynolds stress, i.e. radial transport of poloidal momentum, drives the poloidal flow. When the ensemble average is taken over time $\langle \cdot \rangle_t$, it states that a local Reynolds stress gradient drives a local (stationary) flow. For now only this temporal mean $\bar{R}(\mathbf{r}) = \langle R(\mathbf{r}, t) \rangle_t$ of the local Reynolds stress at the position $\mathbf{r}$,

$$R(\mathbf{r}, t) = \bar{R}(\mathbf{r}) + \tilde{R}(\mathbf{r}, t) , \tag{8.1}$$

is considered. For a non-isotropic, non axis-symmetric velocity distribution, reflected in a correlation of radial and poloidal velocity, the mean Reynolds

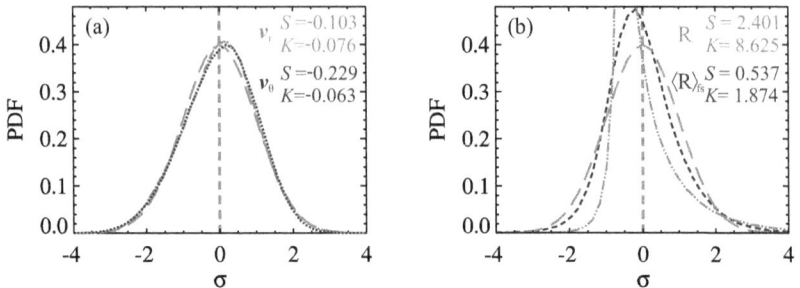

Figure 8.1: Probability distribution functions (PDF) of the velocity components (a) and the resulting Reynolds stress (b) (#9985). Radial $\tilde{v}_r$ (red dash-dot) and poloidal velocity fluctuations $\tilde{v}_\theta$ (blue dots) are close to a Gaussian distribution (grey long dashes), whereas the resulting local Reynolds stress R (red dash-dot-dot) and flux surface averaged Reynolds stress $\langle R \rangle_{fs}$ (blue dashes) are strongly non-Gaussian. Skewness $\mathcal{S}$ and kurtosis $\mathcal{K}$ are shown for each distribution.

stress value is non-zero. The influence of the magnetic field geometry on the phase relation between the velocity components is analysed.

8.1.1 Geometry dependence

With the poloidal probe array, the velocity components are measured in the complete edge region of the confined plasma in a helium discharge at low magnetic field (#9985). For the angle $\theta \approx 0.4\pi$ figure 8.1 shows the probability distribution functions (PDF) of the velocity components $\tilde{v}_{r,\theta}$, the resulting local Reynolds stress R, and the flux surface averaged Reynolds stress $\langle R \rangle_{fs}$. Skewness $\mathcal{S}$ and kurtosis $\mathcal{K}$ of each distribution are given in the respective figure. In comparison with the reference Gaussian distribution (grey long dashed line) both radial and poloidal velocities exhibit a near Gaussian statistics. Due to the non-linearity, the local Reynolds stress has a very high kurtosis and is positively skewed. A positive skewness implies that on average events with outward-going transport and positive poloidal velocity or inward-going negative velocity events dominate the Reynolds stress at this position. Further, the moments of the PDF are an indication of an intermittent or bursty momentum transport, which was found in other experiments as well [205]. Also after the flux surface average $\langle \cdot \rangle_{fs}$ has been taken the Reynolds stress distribution is distinctly non-Gaussian with a skewness of $\mathcal{S} = 0.537$ and a kurtosis of $\mathcal{K} = 1.874$.

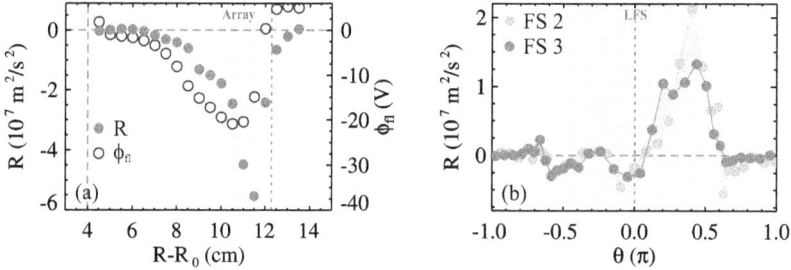

Figure 8.2: The radially resolved local Reynolds stress $\bar{R}(R-R_0)$ (red filled dots) as well as the floating potential profile ϕ_{fl} (blue open dots) are shown in (a) (#9821). The Reynolds stress is strongest in the edge region, where the poloidal probe array is positioned (light grey area). On the right hand side (b), the poloidal profile of the Reynolds stress $\bar{R}(\theta)$ on the two neighbouring flux surfaces (FS 2 and FS 3) is presented (#9985). Non-zero values of the Reynolds stress are restricted to the low field side (LFS) with a strong poloidal asymmetry.

As the velocity fluctuations strongly depend on the spatial position, also the Reynolds stress is assumed to show a spatial dependence. Using the 5-pin probe, the radial potential and Reynolds stress profile is measured in the midplane from the plasma centre ($R-R_0 = 4$ cm) to the scrape-off layer (Fig. 8.2 (a)). The floating potential ϕ_{fl} (blue open circles) has a minimum in the edge region and is zero at the separatrix ($R-R_0 = 12.3$ cm). A similar structure is found for the mean Reynolds stress $\bar{R}$ (red filled circles) which is here always negative with a sharp minimum at $R - R_0 = 11.5$ cm. It should be stressed that the values are not calculated from the mean values of the potentials, but rather from the temporal mean taken of the product of the velocity fluctuations calculated according to equation (5.10). Similar experiments in a linear plasma device (CSDX) showed that the divergence of the radial mean Reynolds stress drives an azimuthally symmetric flow [206, 207]. However, in this experiment a poloidally symmetric Reynolds stress profile cannot be assumed and the angular dependence is studied in the following.

The poloidal probe array covers the extreme regions of the Reynolds stress profile and its radial position is marked with a grey box in figure 8.2 (a). As mentioned before, the flux surfaces at this toroidal position have similar geometrical properties as the field lines in a tokamak (cf. Fig. 5.2). In figure 8.2 (b) the poloidal mean Reynolds stress profile $\bar{R}(\theta)$ for both flux

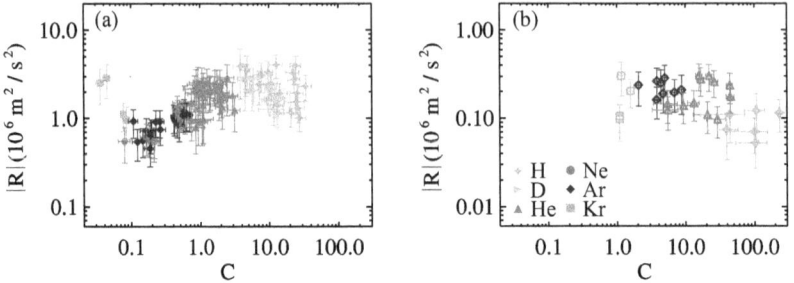

Figure 8.3: Scaling of the mean Reynolds stress $|\bar{R}|$ for low (a) and high magnetic field (b). The poloidally averaged absolute values of the Reynolds stress, measured with the probe array, are scaled with collisionality.

surfaces is shown. The poloidal dependence is similar for both flux surfaces, whereas differences, marked as coloured region in the figure, measure the radial Reynolds stress gradient. They indicate flow drive and are mainly found on the low field side (LFS). A pronounced maximum is visible above the midplane ($\theta \approx 0.4\pi$). In previous investigations a similar poloidal dependence has been found for the radial cross-field transport $\Gamma = \langle \tilde{v}_r \tilde{n} \rangle$ [91, 208]. These analyses have shown that the turbulent transport is peaked in regions where the normal curvature κ_n is negative and the geodesic curvature κ_g is positive. This is in line with theoretical studies which predict a ballooning of fluctuation amplitudes for $\kappa_n < 0$ [209, 210] and it shows an additional influence by the geodesic curvature. Since the Reynolds stress $\mathcal{R} = \langle \tilde{v}_r \tilde{v}_\theta \rangle$ can be seen as radial transport of poloidal momentum and it also directly depends on the underlying drift-wave turbulence, a comparable influence of the curvature terms can be assumed. With the gyrofluid code GEMR the asymmetric poloidal structure of the Reynolds stress and the turbulent transport could be recovered in simulations for TJ-K parameter [211]. However, further analysis of the simulations suggest that the magnetic shear should have a major influence.

The experimental possibilities to systematically study the influence of the magnetic field are however limited. A continuous variation of the magnetic field parameters, by changing the current ratio, cannot be performed since this would change the position of the flux surfaces. But through a magnetic field reversal the poloidal course is mirrored at the midplane which does only affect the parameters with asymmetric poloidal profiles. In figure 8.4, and similarly in figure 8.5, the poloidal dependences of normal curvature κ_n and geodesic curvature κ_g (c,d) as well as integrated magnetic shear Λ and local magnetic shear S (e,f) are shown for the toroidal position of port O2 for standard $(+\mathbf{B})$ and reversed magnetic field $(-\mathbf{B})$. The normal curvature (blue solid line) is up-down symmetric with negative values, i.e. 'bad' curvature, on the outboard side, whereas the geodesic curvature (green dashed line) changes sign at the midplane and has a sinusoidal form with minimum and maximum at bottom and top, i.e. $\theta \approx \mp\pi/2$, respectively. Both, integrated magnetic shear (red solid line) and local magnetic shear (magenta dashed line), have a more complicated poloidal dependence and exhibit extreme values at $\theta \approx \pm 0.6\pi$. With a field reversal the geodesic curvature and the integrated magnetic shear change sign whereas normal curvature and local magnetic shear stay the same. An influence of all magnetic field parameters on the Reynolds stress has to be assumed. It turns out that the combination of normal curvature, geodesic curvature, and integrated local magnetic shear determines the growth rate of the drift-wave instability when the full curvature vector is included in the calculations [101],

$$\gamma_{\mathrm{DW}} \propto -\kappa_n + \kappa_g \left(\frac{k_s}{k_\alpha} + \Lambda \right) , \tag{8.2}$$

with the radial k_s and poloidal wavenumber k_α in terms of field aligned coordinates. Since the turbulent fluctuations enter quadratically in the Reynolds stress, the poloidal variation of the growth rate might regulate the Reynolds stress amplitude. On the other hand the turbulent structures have a finite spatial extent which is affected by the magnetic shear (see Fig. 5.3). The importance of each effect is difficult to estimate, however, the expected influences are the following:

- The normal curvature κ_n leads to a destabilisation of density and potential perturbations when it is parallel to the magnetic field gradient, i.e. $\kappa_n < 0$ [212] (cf. Chap. 2.2). This results in higher fluctuation levels on the LFS of toroidal experiments. Thus a ballooning envelope is expected with a maximum at the outboard midplane $(\theta \approx 0 \cdot \pi)$.

- In context of the growth rate the geodesic curvature κ_g has to be considered in combination with the integrated local magnetic shear Λ. Without magnetic shear, positive values of the geodesic curvature would increase the growth rate which would be in the experiment above the midplane for $+\mathbf{B}$ and below for $-\mathbf{B}$. The strength depends on the ratio of radial k_s to poloidal wavenumber k_α.

- As the integrated local magnetic shear Λ enters the growth rate as a product with the geodesic curvature, its effect does not change with magnetic field reversal. At the toroidal measurement position it leads to an increase of the growth rate above and below the midplane on the outboard side. Since the drift waves are mostly field aligned, a integrated local shear of neighbouring field lines would also be transferred to the turbulent structures. In the coordinate system of the experiment, a positive integrated local magnetic shear would entail positive Reynolds stress and vice versa.

- The role of the local magnetic shear S is more complicated (cf. Eq. (5.3)). From simulations [213] a stabilising effect for drift waves is suspected which might decrease the growth rate and thus lower the Reynolds stress.

In [101] also the influence of the heating position on the growth rate is discussed, which could lead to additional asymmetries. The overall Reynolds stress level, as the fluctuation amplitudes, depends on the background plasma parameters. This can be seen with figure 8.3 which shows the absolute values of the Reynolds stress $|\bar{R}|$ averaged poloidally. An increasing trend, as for the fluctuation level (cf. 6.3.1), is found for lower collisionality, at least in the low magnetic field case (a).

To reduce the poloidal variations not originating from the magnetic field geometry, the poloidal Reynolds stress profiles $\bar{R}(\theta)$ of several measurements at different control parameters are averaged. Each measurement is thereby normalised to the overall Reynolds stress level $\sum |\bar{R}|$. The poloidal dependency is shown in figure 8.4 for standard magnetic field direction $+\mathbf{B}$ (left) and reversed magnetic field $-\mathbf{B}$ (right) for helium discharges at low magnetic field.

The averaged Reynolds stress profile with standard field direction (a) shows a similar dependency as the single measurement (Fig. 8.2 (b)), with only little variation implied by the small error values (95 % confidence interval). The position of the maximum falls into the region with negative normal and positive geodesic curvature (marked as grey region) which supports the hypothesis that the curvature influenced growth rate determines the Reynolds stress. Poloidal positions with high magnetic shear, either local or integrated, are indicated as hatched area.[1] Noticeable Reynolds stress values are found for the angle $\theta = 0 \cdot \pi$ where the local magnetic shear is positive, but

---

[1] The inboard side is omitted due to the generally small Reynolds stress amplitudes.

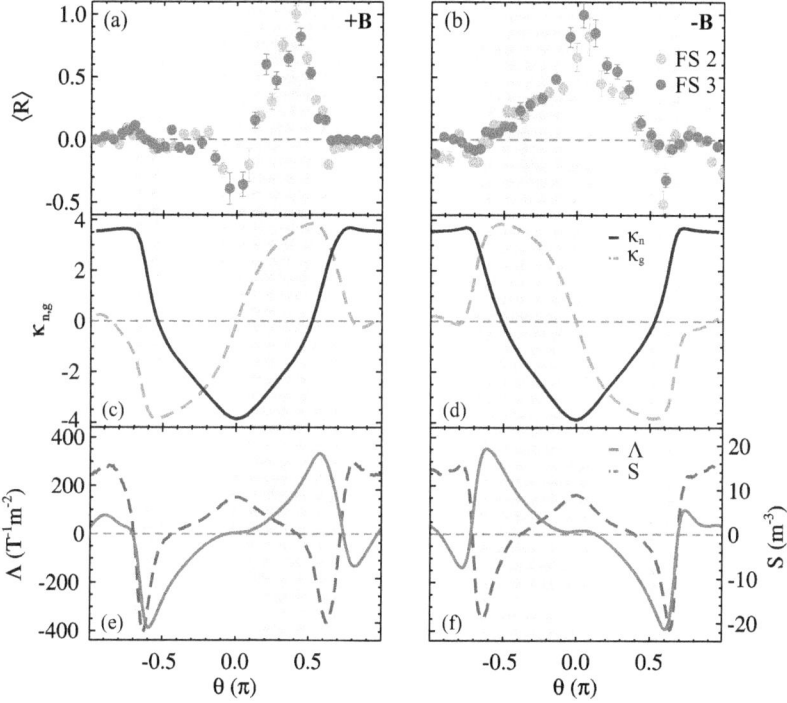

Figure 8.4: Poloidal dependence of the mean Reynolds stress $\bar{R}(\theta)$ for low magnetic field helium plasmas. On the left hand side (a) the configuration with standard magnetic field direction is shown, whereas for the measurements on the right (b) the magnetic field direction is reversed. In figure (c,d) the corresponding poloidal trend of the curvature components can be seen, with normal curvature κ_n (blue solid line) and geodesic curvature κ_g (green dashed line). Below (e,f) the dependency of the integrated magnetic shear Λ (red solid line) and the local magnetic shear S (magenta dashed line) is shown. (see text for further information)

the integrated magnetic shear is zero. However, in the other two regions, i.e. $\theta \approx \pm 0.6\pi$, where the local magnetic shear is negative, only small values are found and the Reynolds stress seems to have opposite sign. As these positions are already on the inboard side, high Reynolds stress values are not necessarily suspected as the normal curvature is positive and fluctuation levels are therefore low.

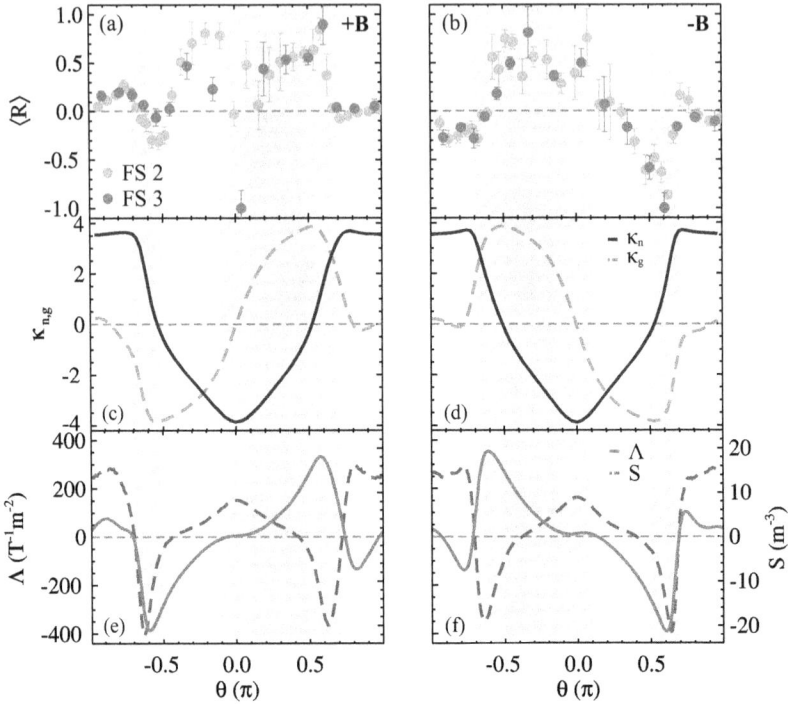

Figure 8.5: Same representation as in figure 8.4 but for the case of high magnetic field. The poloidal Reynolds stress distribution is similar, but the influence of the magnetic shear seems to be stronger.

With a magnetic field reversal the geodesic curvature and the integrated magnetic shear change sign. The resulting poloidal Reynolds stress profile is shown in figure 8.4 (b). Also below the midplane ($\kappa_n < 0$, $\kappa_g > 0$) (grey region) the Reynolds stress has now significant values, where, however, the Reynolds stress is generally large for negative normal curvature. More noticeable is the course at the positions of large magnetic shear (hatched area). The Reynolds stress has now a sharp maximum at $\theta = 0 \cdot \pi$. Also in the other two regions the sign of the Reynolds stress seems to be switched although the influence of the magnetic shear is again marginal.

For high magnetic field the field configuration stays the same but the plasma turbulence, especially the structure size (Chap. 6.2.3), changes strongly. The found dependencies (Fig. 8.5) are in principle the same as in the case of low magnetic field whereas stronger radial variations are observed. The influence of the geodesic curvature seems to shift the Reynolds stress from above to below the midplane when the field is reversed. But due to the superposition of multiple effects, the data is not conclusive. However, the effect of the magnetic shear can be identified more clearly. At the outboard midplane ($\theta = 0 \cdot \pi$) the local magnetic shear leads to negative Reynolds stress for the standard field direction $+\mathbf{B}$ and to positive Reynolds stress in the reversed case $-\mathbf{B}$. The poloidal dependency of the Reynolds stress seems to follow the profile of the integrated magnetic shear with the extreme values at $\theta \approx \pm 0.6\pi$ and opposite sign. Additionally, it might be speculated that local and integrated magnetic shear interact constructively at the top ($\theta \approx +0.6\pi$) and destructively at the bottom ($\theta \approx -0.6\pi$) due to the different sign dependence. However, it remains complicated to distinguish between the different influences of curvature and shear induced Reynolds stress. Therefore, measurements at other toroidal positions would be desirable, but a complete separation of the different influences is not possible in the TJ-K device.

8.1.2 Coherence of velocity components

Like in the case of turbulent particle transport Γ where a positive correlation of radial velocity $\tilde{v}_r$ and density fluctuation $\tilde{n}$ result in an outward cross-field transport (see Chap. 2.2), a correlation of radial and poloidal velocity leads to non-zero Reynolds stress. Written with the cross-power spectrum (cf. Eq. (2.20)) this is[2]

$$\bar{R}(\theta) = \sum_f \gamma_{v_r,v_\theta}(\theta, f) \sqrt{S_{v_r}(\theta, f) S_{v_\theta}(\theta, f)} \cos(\alpha_{v_r,v_\theta}(\theta, f)) \, . \qquad (8.3)$$

$\gamma_{v_r,v_\theta}(f)$ is the cross-coherence between the velocity fluctuations, $S_{v_r}(f)$ and $S_{v_\theta}(f)$ are the respective auto-power spectra, and $\alpha_{v_r,v_\theta}(f)$ is the cross-phase spectrum.

---

[2]In the same way, this can be formulated in wavenumber space.

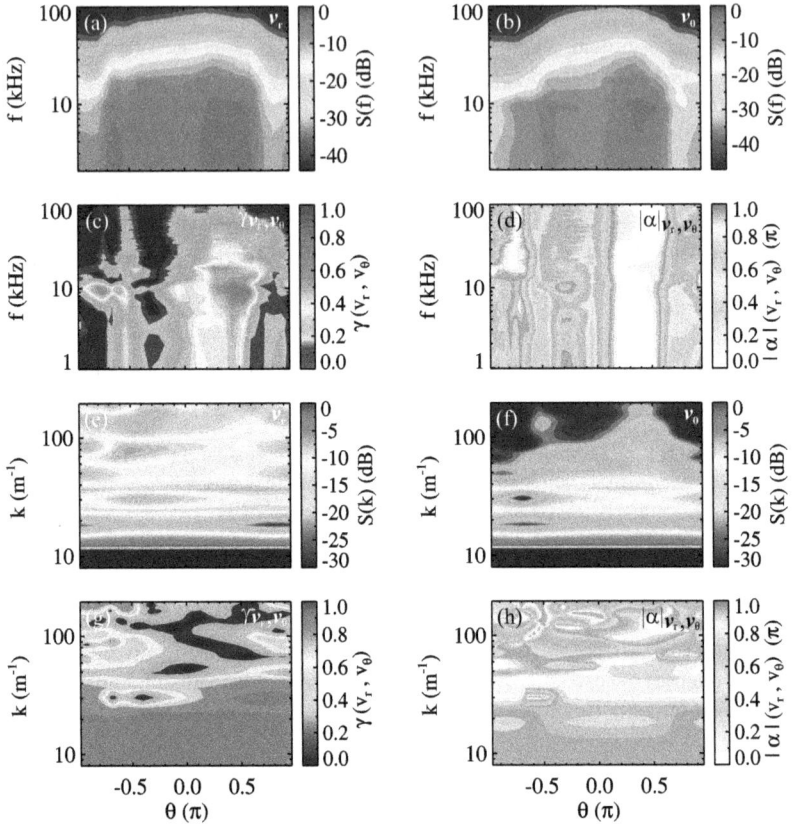

Figure 8.6: Poloidally resolved frequency spectra (a,b) and wavenumber spectra (e,f) of the radial v_r and poloidal v_θ velocity component, cross-coherence γ_{v_r,v_θ} (c,g) and cross-phase spectra $|\alpha|_{v_r,v_\theta}$ (d,h) for low magnetic field with standard field direction in helium (#9985). The contribution of each scale can be spatially (poloidally) resolved by means of a wavelet transformation. In regions with strong mean Reynolds stress ($\theta \approx 0.4\pi$) high coherence and a phase close to zero is found.

In figure 8.6 the poloidally resolved spectra are shown in frequency (a–d)[3] and wavenumber space (e–h), calculated with a wavelet transformation (Chap. 4.3.2). The frequency spectra of the velocity components (a,b) are broad for the whole poloidal circumference with the typical ballooning envelope. Especially above the midplane ($0 < \theta < \pi/2$) strong contributions are found for the poloidal velocity component v_θ at frequencies around $10\,\text{kHz}$. Also the wavenumber spectra (c,d) show a similar poloidal distribution with the asymmetry in the poloidal velocity and significant contributions up to the smallest scales. In the region of the pronounced poloidal Reynolds stress maximum ($\theta \approx 0.4\pi$), high coherence levels (c) are found for frequencies above $10\,\text{kHz}$, which are associated with the dominant drift wave structures. Viewed in wavenumber space (e) the coherence is relatively high for most wavenumbers and the poloidal dependency is not as clear as in the frequency domain.[4] This illustrates that for the actual mean Reynolds stress both coherence and phase of the fluctuating velocity components have to be considered. The phase spectrum (d) is close to zero for all frequencies at the position of the Reynolds stress maximum ($\theta \approx 0.4\pi$) and the same can be observed for the spatial scales (h). However, the main poloidal variation in the wavenumber resolved phase spectrum (h) lies in the scales above $100\,\text{m}^{-1}$.

For a further investigation the relation between the velocity components is compared at three distinct poloidal positions (Fig. 8.7). The bivariant probability distribution function (2D-PDF) as well as the coherence and the phase spectra are shown at the inboard side (a,d,g), the outboard side (b,e,h), and for the maximum Reynolds stress region (c,f,i). In the 2D-PDF the connection between the velocity components is directly visible, where a correlation or an anticorrelation manifests itself in an anisotropic velocity distribution. For a better comparison the $1/e$ level of the reference Gauss is plotted as white circle. On the inboard side ($\theta \approx -\pi$) the velocity distribution is isotropic, which results in a Reynolds stress value near zero. For the midrange frequencies and nearly all wavenumbers the coherence is significant, but the corresponding phase changes between 0.2π and 0.6π, or 0π and π respectively, which in total leaves no strong mean contribution. The situation changes when the positions on the outboard side are considered.

---

[3] A boxcar average with the two neighbouring points is applied in poloidal direction to get a clearer picture.

[4] This is due to the nature of the wavelet transformation where the scales are blurred and the spatial resolution is relatively poor.

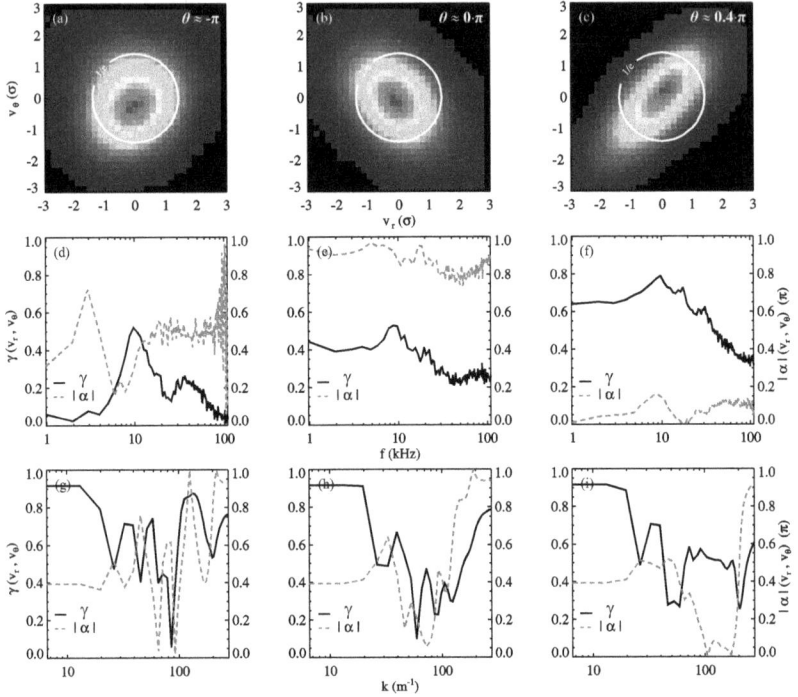

Figure 8.7: Connection between radial and poloidal velocity components for three different poloidal locations (#9985). In the upper row (a,b,c) the bivariant probability distribution functions can be seen. The $1/e$ level of the reference Gaussian is drawn as solid white line. The lower rows (d–i) show the corresponding cross-coherence γ_{v_r,v_θ} as well as the absolute value of the cross-phase spectrum $|\alpha|_{v_r,v_\theta}$ between the velocity components. At the outboard midplane the Reynolds stress is negative, which is a consequence of the anisotropic velocity distribution (b) and the phase shift of π between the two velocity components. At $\theta \approx 0.4\pi$ a phase shift near zero (f,i) results in a positive Reynolds stress.

144

The 2D-PDFs are strongly anisotropic pointing to a high resulting mean Reynolds stress and explain the values of skewness and kurtosis found before (Fig. 8.1 (b) for $\theta \approx 0.4\pi$). For the angle $\theta \approx 0 \cdot \pi$ (Fig. 8.7 (e)) the absolute value of the phase is near π for frequencies with high coherence, which constitutes the negative correlation of the velocity components. The same is true for wavenumbers above $100\,\mathrm{m}^{-1}$ where the phase shift is constantly large. In contrast a positive Reynolds stress, as for $\theta \approx 0.4\pi$, is reflected by zero phase shift (f). Also here high wavenumbers seem to dictate the sign of the Reynolds stress. For the zonal flow the fluctuating Reynolds stress is important which will be analyses for its spatial variation in the next section.

8.2 Reynolds stress dynamics

So far the temporal mean of the Reynolds stress $\bar{R}(\mathbf{r})$ was analysed in detail as for now the fluctuating part will be considered,

$$R(\mathbf{r}, t) = \tilde{R}(\mathbf{r}, t) \ . \tag{8.4}$$

This means that either the mean Reynolds stress is not shown (i.e. $f \neq 0$) or that it is explicitly subtracted for the analysis (i.e. $\langle R(\mathbf{r}, t) \rangle_t = 0$). Similar to the previous section, the spatial Reynolds stress signal power distribution will be shown first (Sect. 8.2.1), and then in section 8.2.2 the connection to the zonal flow will be made. In the driving mechanism of the zonal flow the density-potential coupling plays an important role (see Chap. 3.2) which is analysed by means of the pseudo-Reynolds stress in section 8.2.3.

8.2.1 Higher order moments

Figure 8.8 (a) shows the poloidally resolved auto-power frequency spectrum of the Reynolds stress $S_R(\theta, f)$, which in total is proportional to the standard deviation $\sigma \propto \sum_{f \neq 0} S_R(\theta, f)$. The same discharge as in the section before is considered and also here a boxcar average with the neighbouring points is applied. In comparison with the poloidal profile of the velocity components (Fig. 8.6 (a,b)), the Reynolds stress fluctuations exhibit an even stronger inboard-outboard asymmetry, and again the maximal amplitudes can be found above the midplane. In addition, this asymmetry is reflected in the frequency range that contributes most to the Reynolds stress. The poloidal structure resembles the poloidal profile of the mean Reynolds stress (Fig. 8.2 (b)), although it is more continuous, and even the magnetic shear

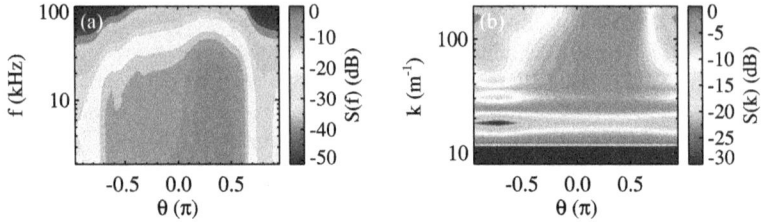

Figure 8.8: Poloidally resolved Reynolds stress spectra for a low magnetic field (standard direction) helium discharge (#9985). Figure (a) shows the Reynolds stress in frequency space and figure (b) in wavenumber space. The Reynolds stress is strongest on the outboard side and asymmetric to the upper half where smaller scales dominate.

influences seem visible. The discussion of the influence of the magnetic field parameter is essentially the same as in chapter 8.1.1.

Using a wavelet transformation (Chap. 4.3.2) the contribution of each scale can be spatially (poloidally) resolved. The wavenumber spectrum $S_R(\theta, k)$ (Fig. 8.8 (b)) shows that especially small scale structures dominate the Reynolds stress fluctuations where the maximum is located above the midplane. This is reminiscent of an earlier study on TJ-K [34], supported by computational results [113], where it was shown that the zonal flow is predominantly driven by the smaller scales.

Next, the collisionality scaling of the moments of the turbulent Reynolds stress is investigated. The dependency of the standard deviation, equivalent to a fluctuation level, on the collisionality is shown in figure 8.9 for low and high magnetic field. The trend resembles the one of the fluctuation level of density and potential (Chap. 6.3.1) and is similar to the scaling of the mean Reynolds stress $\bar{R}$. Thus it appears that when the fluctuations in the potential get more violent also the Reynolds stress (mean and fluctuations) increases. This could also imply that the zonal flow drive increases the same way as the Reynolds stress gradients get steeper. However, the relative zonal flow power increases for decreasing collisionality (cf. Chap. 7.3.2).

As in the case of the flux surface averaged potential, higher order moments can capture characteristics of the turbulent dynamics. The collisionality scalings of skewness and kurtosis are shown in figures 8.10 (a,b) and (c,d), respectively, for low and high magnetic field. To obtain a representative value for each discharge the poloidal average is taken over all measurement

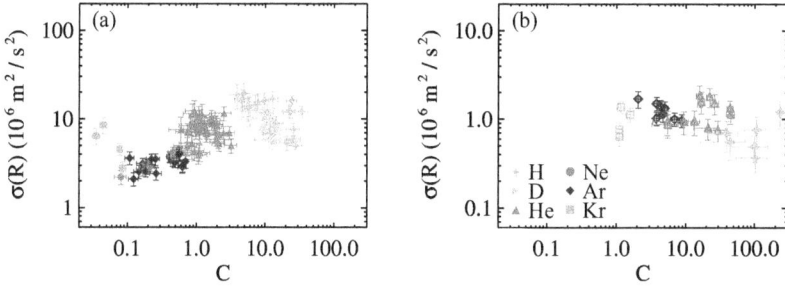

Figure 8.9: Scaling of the standard deviation σ of the Reynolds stress with collisionality. The average over all positions of the Reynolds stress array is calculated,
which is equivalent to the overall fluctuation level of the Reynolds stress. The
scaling for low magnetic field is shown in (a) and for high magnetic field in (b).

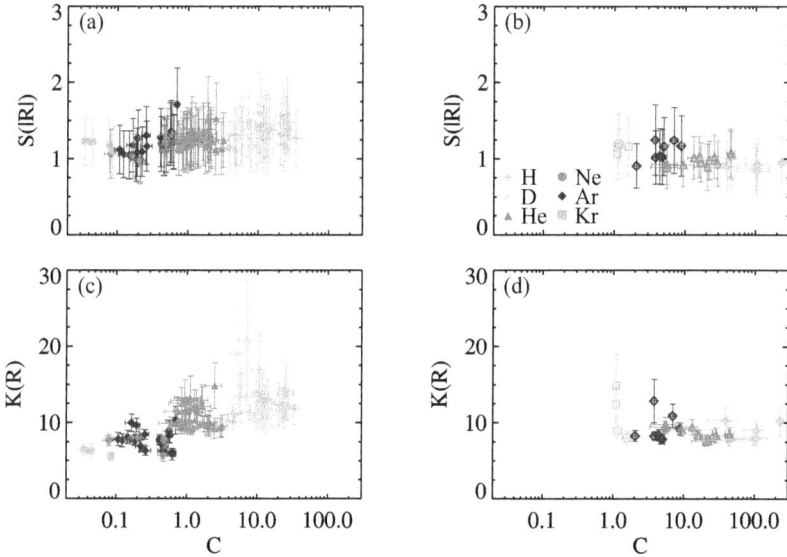

Figure 8.10: Collisional scaling of higher order moments of the turbulent Reynolds
stress, i.e. skewness $\mathcal{S}$ (a,b) and kurtosis $\mathcal{K}$ (c,d), for low and high magnetic field,
respectively. The skewness is constant whereas the kurtosis increases when viewed
for a wide range of collisionality.

points. The skewness, where the absolute value is shown, stagnates and has similar values for all measurements. This means that the Reynolds stress fluctuations are balanced and neither positive nor negative fluctuations become dominant with changing parameters. Compared with the distribution of the zonal potential, this fits to the equipartition of the zonal flow events (Chap. 7.1). The kurtosis, on the other hand, is generally high and shows an increase with collisionality, at least for the collisionality range covered by the low magnetic field measurements (c). As the kurtosis quantifies the importance of the wings in comparison to the normal distribution, it is a sign for an increased intermittency. A general intermittent behaviour is predicted for the vorticity but an increase of the intermittency level is known from the density which gets intermittent as it becomes a passively advected scalar [135, 214, 215]. The theoretically expected scaling for the Reynolds stress is, however, unclear and simulations are needed to resolve the role of collisionality for the Reynolds stress.

8.2.2 Zonal flow drive

In a next step, the connection to the zonal flow will be drawn. As done in chapter 7, the dynamics around the zonal flow occurrence is obtained by applying the conditional averaging technique (Chap. 4.6). The zonal potential $\langle \tilde{\phi} \rangle_{\mathrm{fs}}$, approximated as poloidally averaged signal from the floating potential of the probes on the 3rd flux surface (FS 3), is used as trigger signal with the condition $+2\sigma$. The subwindows are centred on the respective maximum, so that $\tau = 0\,\mu\mathrm{s}$ marks the position of the maximal zonal potential. In total 896 realisations are used for the ensemble average. Figure 8.11 shows the averaged dynamics in wavenumber space of the Reynolds stress on flux surface FS 2 (a) and FS 3 (b) as well as the radial Reynolds stress gradient (c). The Reynolds stress $\tilde{\mathrm{R}}(k, \tau)$ on both flux surfaces fluctuates periodically with main contributions at wavenumbers $k > 70\,\mathrm{m}^{-1}$, getting more intense before the zonal potential has its maximum. The oscillation is transferred to the Reynolds stress gradient $\partial_r \tilde{\mathrm{R}}(k, \tau)$ which is strongest at even smaller scales.

For the same time as in figure 8.11, the Reynolds stress evolution $\tilde{\mathrm{R}}(\theta, \tau)$ is shown in real space as coloured contour plot in figure 8.12 (a). The connected local radial gradient of the Reynolds stress $\partial_r \tilde{\mathrm{R}}(\theta, \tau)$ is overlaid as contour lines, where continuous and dashed lines show negative and positive gradients, respectively. Although the fluctuations in both quantities are small in scale, it is clear that strong contributions of turbulent Reynolds

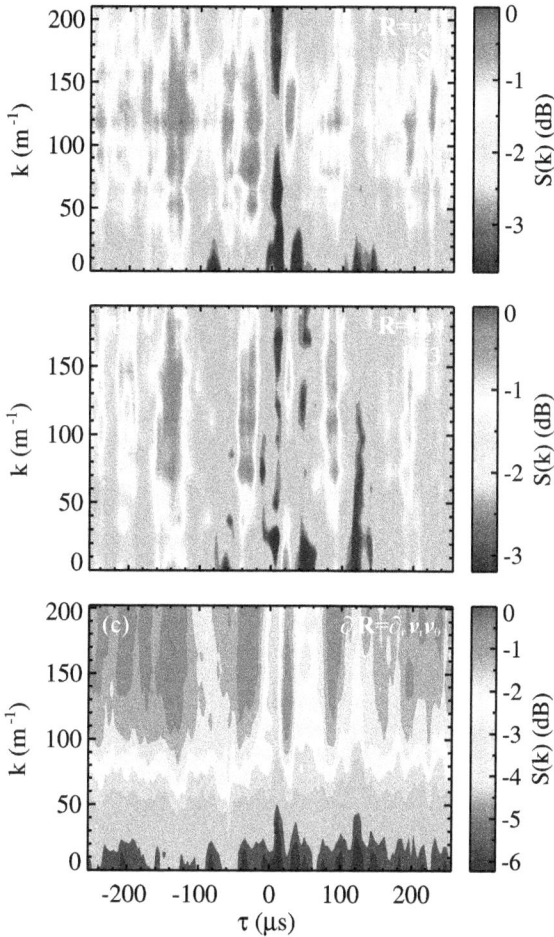

Figure 8.11: Temporal evolution of the Reynolds stress and Reynolds stress gradient in wavenumber space around the zonal flow occurrence. The Reynolds stress $\tilde{R}(k, \tau)$ on flux surface FS 2 (a) and FS 3 (b) show a primary oscillation with main contributions at $k > 70\,\text{m}^{-1}$. This is transferred to the Reynolds stress gradient $\partial_r \tilde{R}(k, \tau)$ (c) which is strongest at even smaller scales.

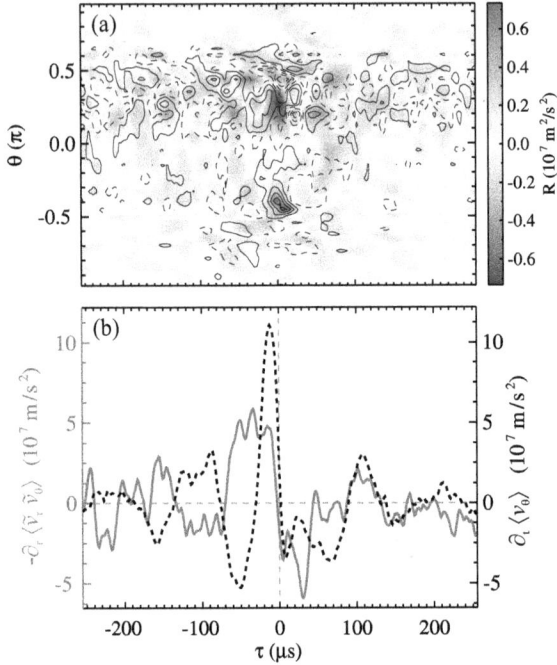

Figure 8.12: Conditional averaged dynamics of Reynolds stress drive and zonal flow response. For the time evolution around the trigger time point the Reynolds stress $\tilde{R}(\theta, \tau)$ (filled contour) and the local Reynolds stress gradient $\partial_r \tilde{R}(\theta, \tau)$ (contour lines) are plotted in (a). Both Reynolds stress and Reynolds stress gradient are strong on the outboard side. For the same timescale the radial gradient of the flux surface averaged Reynolds stress $-\partial_r \langle \tilde{v}_r \tilde{v}_\theta \rangle_{\mathrm{fs}}$ (red solid) and the acceleration of the poloidal flow $\partial_t \langle v_\theta \rangle_{\mathrm{fs}}$ (black dashed) are shown below (b).

stress and its local gradient are restricted to the outboard side of the plasma. Beyond that, the up-down asymmetry, already revealed by the spectra (cf. Fig. 8.8), can be detected for the whole time evolution around the zonal flow. Shortly before the trigger condition is reached ($\tau = 0\,\mu s$) the Reynolds stress gradient shows contributions also on the lower side ($\theta \approx -0.4\pi$). But for the zonal flow the flux surface averaged quantities have to be considered since they are the driving force of the poloidal flow (Eq. (3.17)).[5] The zonally averaged terms of the drive equation are shown in the lower part of figure 8.12 for the same time scale. With a red solid line the time evolution of the Reynolds stress drive $-\partial_r \langle \tilde{v}_r \tilde{v}_\theta \rangle_{\mathrm{fs}}$ is displayed, whereas the acceleration of the poloidal flow $\partial_t \langle v_\theta \rangle_{\mathrm{fs}}$ is drawn in dashed black. Similar to the poloidally resolved picture, the Reynolds stress drive fluctuates fast as compared to the poloidal flow. Shortly before the flow gets maximal the Reynolds stress drive is strong and reaches comparable absolute values. Both pictures together, spatially resolved and flux surface averaged, show that the zonal flow is driven by the Reynolds stress, and this drive turns out to be poloidally localised.

The influence of the poloidal shear flow on the ambient turbulence can be examined on the coherence and phase spectra of radial and poloidal velocity components where the time resolution is again obtained by a conditional average. Also here the absolute phase is plotted since it is the determining quantity for the Reynolds stress. Figure 8.13 (a,c) show the time evolution of both quantities around the zonal flow occurrence. The variation in the coherence $\gamma_{v_r,v_\theta}(k)$ is relatively small and considerable values are found for almost all wavenumbers in the range $k > 70\,\mathrm{m}^{-1}$. On the other hand, the cross-phase $|\alpha_{v_r,v_\theta}|(k)$ shows significant variations around the zonal flow occurrence, especially for high wavenumbers. For a better illustration of the development of the phase relation between the velocity components, the coherence and phase spectra at different time points from $\tau = -80\,\mu s$ to $0\,\mu s$ are shown in (b) and (d), respectively. The data is smoothed with the neighbouring points for a clearer illustration. As the maximum of the zonal flow is reached ($\tau = 0\,\mu s$) the deviation to the temporal mean phase (black dashed line) increases which, again, is highest for the smallest scales. To get an impression of the evolution of the overall cross-phase, the spectral averaged deviation of the phase is plotted in figure 8.14. The solid blue

---

[5]The damping of the zonal flow by the ion viscosity (Chap. 3.4.1), which is small due to the high collision rates, and other damping mechanisms as, e.g. the geodesic transfer effect (Chap. 3.4.2) [113, 216, 217]), are not further considered in the present investigation.

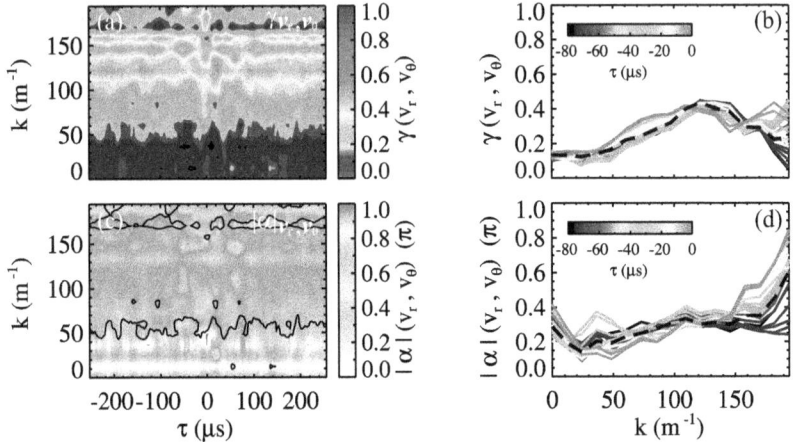

Figure 8.13: Time resolved coherence and phase spectrum of the velocity components on FS 3. The coherence $\gamma_{v_r,v_\theta}(k)$ (a) is overall constant with only little variation around the zonal flow maximum ($\tau = 0\,\mu s$). Similarly, the absolute value of the cross-phase $|\alpha_{v_r,v_\theta}|(k)$ (c) has major deviations from the mean mostly when the zonal flow is strong. Regions of considerable coherence ($\gamma \geq 0.2$) are shown as black solid contour line. As illustration of the influence of the zonal flow on the relation between the velocity components, the development of coherence and phase from $\tau = -80\,\mu s$ to $0\,\mu s$ is shown in (b) and (d), respectively. For a clearer visualisation the data has been smoothed. The black dashed line marks the mean level.

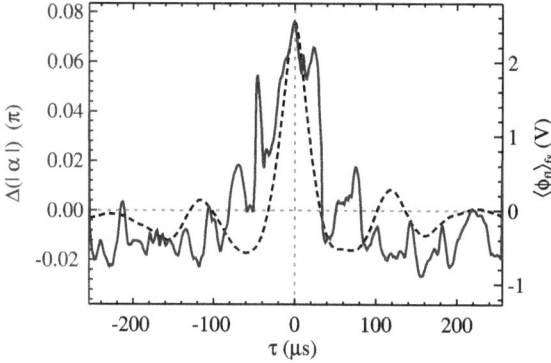

Figure 8.14: Development of the mean absolute deviation of the cross-phase $\Delta(|\alpha|)$ between radial and poloidal velocity component on flux surface FS 3 around the zonal flow occurrence. At around $\tau = -80\,\mu s$ the phase (solid blue line) starts to deviate from the mean and gets maximal where the flux surface averaged potential (black dashed line) has its maximum.

line shows the mean absolute deviation of the cross-phase and the black dashed line is the flux surface averaged potential. At around $\tau = -80\,\mu s$ the phase starts to deviate from the mean and gets maximal at the trigger time point. The shear flow does indeed alter the phase relation between the velocity components where this shearing seems most effective for smaller scales. This corresponds to the straining-out mechanism and, equivalently, to the manifold shrinking (Chap. 3.2), where the small scale structures are forced by the shear to couple to the zonal flow (see also Chap. 9).

In the example shown in figure 8.12 the absolute value of the Reynolds stress drive does not exceed the actual zonal flow response. However, they are of the same magnitude and for other measurements (Fig. 8.15) the drive is large enough to quantitatively explain the acceleration of the flow. There are multiple reasons why the measured drive can be lower than the acceleration. First of all, its not the perpendicular Reynolds stress which is measured with the array and, therefore, the Reynolds stress is lower than the real poloidal Reynolds stress. Furthermore, the array does only cover a relatively small area in the edge of the confined region and the radial gradient of the flux surface averaged Reynolds stress is approximated with only two points. Therefore, the Reynolds stress gradient could be underestimated.

153

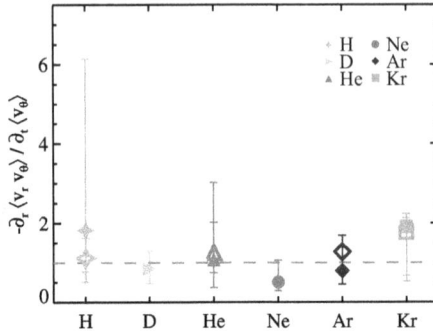

Figure 8.15: Ratio of maximal Reynolds stress drive and zonal flow response, $-\partial_r \langle \tilde{v}_r \tilde{v}_\theta \rangle_{\mathrm{fs}} / \partial_t \langle v_\theta \rangle_{\mathrm{fs}}$, for different gases. The error bars show the range of values and the symbol indicates the median value. Low magnetic field measurements are shown with full symbols and open symbols stand for measurements at high magnetic field.

8.2.3 Cross-field coupling

To get an estimate for the coupling of density and potential, the so-called pseudo-Reynolds stress, originally introduced to get information about the Reynolds stress from density measurements [218], is calculated. The density is, thereby, treated analogue to the potential field. From equations (2.66) and (5.10), then, it follows that the density-based pseudo-Reynolds stress has to be corrected by terms of at least linear order in the collisionality $\mathcal{O}(C)$ leading to the following relation between Reynolds stress R_ϕ and pseudo-Reynolds stress R_n,

$$R = R_\phi = \left\langle \frac{\left(\tilde{n}^{\theta_{i+1}} - \tilde{n}^{\theta_i} \right) \left(\tilde{n}^{r_{i+1}} - \tilde{n}^{r_i} \right)}{r \Delta\theta \, \Delta r \, B^2} \right\rangle + \mathcal{O}(C)$$
$$= R_n + \mathcal{O}(C) \,. \tag{8.5}$$

With a specific bias setting of the poloidal probe array (4th schema, see Chap. 5.2.3) it is possible to measure Reynolds stress and pseudo-Reynolds stress at the same time over the poloidal circumference with a reduced spatial resolution. Probes measuring ion saturation current alternate with probes on floating potential when going around the circumference. Since Reynolds stress and pseudo-Reynolds stress can be measured on two flux surfaces, the corresponding zonal flow drive $-\partial_r R$ of both quantities is obtained. To

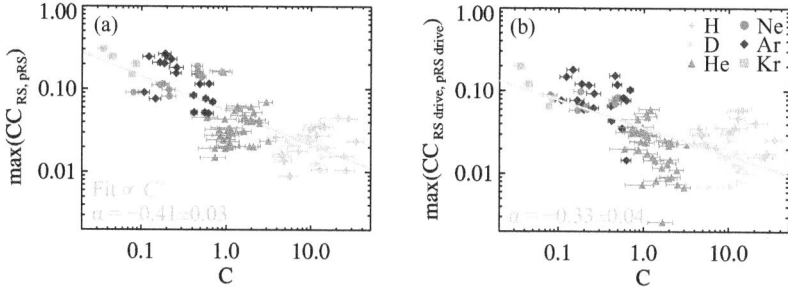

Figure 8.16: Collisionality scaling of the correlation of Reynolds stress and pseudo-Reynolds stress (a) and the respective flow drive (b). The correlation increases for lower collisionality C pointing to an increased density-potential coupling in the adiabatic regime ($C \ll 1$).

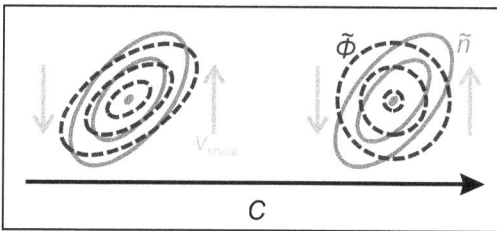

Figure 8.17: The graphic illustrates the change of density (red, solid lines) and potential (dashed, blue lines) coupling with collisionality C. For high collisionality density and potential decouple and the tilt is not transferred to the potential anymore.

obtain a quantitative measure for the similarity of both quantities the cross-correlation of the flux surface averaged Reynolds stress and pseudo-Reynolds stress as well as the respective Reynolds stress drive is calculated from the two time traces of 2^{20} samples for each discharge.

The collisional scaling of the maximal correlation coefficient is shown in figure 8.16 for the Reynolds stress (a) and the Reynolds stress drive (b). Although the correlation values are small, they are significant and show a clear trend. For lower collisionality the correlation between both parameters increases, pointing to an increased coupling between density and potential.

This supports the following picture of the zonal flow drive which is illustrated in figure 8.17 by an eddy in a background shear flow. For an adiabatic electron response (adiabatic regime $C \ll 1$) density and potential act similar, as for the hydrodynamic regime ($C \gg 1$) density and potential act as separate fluids. The spatial shapes of the density (red, solid lines) and the potential perturbation (dashed, blue lines) are shown for low and high collisionality C. A background shear flow (e.g. the zonal flow) tilts the vortex in the density, and, through parallel coupling, also the potential is deformed. In the limit of high collisionality the vortex tilt in the density is not transferred to the potential anymore. In magnetised plasma the potential perturbation leads, via $E \times B$-drift, to vortices perpendicular to the magnetic field. A sheared eddy in the potential has a non-isotropic velocity distribution, giving a non-zero Reynolds stress, and in turn leading to an amplification of the initial shear flow (see Fig. 3.5). With this argumentation it is clear that with an increased collisionality the zonal flow drive is hindered.

8.3 Summary of the chapter

The probe array was used to study the Reynolds stress distribution on the complete poloidal circumference. With the additional radial resolution even the Reynolds stress gradient was estimated which is the important quantity for the zonal flow drive. Using a conditional averaging technique, the evolution of both, poloidally resolved and flux surface averaged Reynolds stress, is analysed in a time window around a zonal flow occurrence. The results for the background Reynolds stress (temporal mean) and its fluctuating component can be summarised:

- The background Reynolds stress shows a ballooning shape with significant values located where the normal magnetic curvature κ_n is negative (outboard side). In spite of the up-down symmetry of the flux surfaces the Reynolds stress maximum is shifted to the region where the geodesic curvature κ_g is positive, i.e. above the midplane. Similarly as for the turbulent cross-field transport Γ [208], this seems to be related to the dependency of the growth rate on both components of the magnetic field line curvature. The average tilt of the turbulent structures is reflected in the anisotropy of the bivariant velocity distribution where especially small scale structures are decisive for the resulting Reynolds stress orientation.

- Additionally to the magnetic curvature also the magnetic shear seems to influence the Reynolds stress which is more pronounced for smaller structure sizes. The integrated magnetic shear as well as the local magnetic shear show an effect on the tilt of the turbulent structures. In the case of the integrated magnetic shear the tilt and, therefore, the sign of the Reynolds stress follows the direction of the shear. For the local magnetic shear this depends on the propagation direction of the turbulent structures.

- The Reynolds stress fluctuations show a similar poloidal dependence as the mean Reynolds stress, suggesting an analogue influence of the background magnetic field. The conditional averaged evolution of the driving term and the poloidal flow illustrates the Reynolds stress drive of the zonal flow, but this drive turns out to be poloidally localised. As the shear flow grows, the cross-phase between the radial and poloidal velocity component starts to deviate from the mean value where the effect seems to be especially pronounced for small structures.

- Reynolds stress and density-based pseudo-Reynolds stress provide a way to determine the cross-field coupling between density and potential. The collisionality scaling of the correlation of both quantities shows that for lower collisionality both quantities are increasingly similar. This demonstrates the transition from a typical plasma response to a neutral fluid behaviour and implies that with decreasing collisionality the zonal flow drive should be more efficient.

Chapter 9

Energy transfer in the drift-wave zonal-flow system

The drift-wave zonal-flow interaction is governed by three-wave coupling and, therefore, best analysed by means of bispectral analysis (Sect. 9.1). This indicates phase coherence of the three interacting modes but it does not show the power transfer between the different modes. To this end, in section 9.2 the nonlinear power transfer function is shown, which gives the direction and amount of the spectral power transfer as it includes the coupling coefficients (see Chap. 4.5). The focus of this work is on the cross-field nature of the plasma turbulence. Therefore, the transfer function considering the cross-coupling between density and potential is calculated using fluctuations of both fields (denoted with the superscript N). This corresponds to the density fluctuation activity, whose transfer function is most sensitive to the adiabaticity parameter [39]. For comparison, also the power transfer of the fluid kinetic energy is calculated (denoted with the superscript V). The collisional scaling of the energy transfer for different modes is shown in section 9.2.2.

9.1 Nonlinear coupling with zonal flow

The degree of nonlinear three-wave coupling is measured by the bispectrum (see Chap. 4.4). The normalised form is the bicoherence

$$b^2(k_1, k_2) := \frac{|\langle \varphi(k_1, t)\varphi(k_2, t)\varphi^*(k_3, t)\rangle|^2}{\langle |\varphi(k_1, t)\varphi(k_2, t)|^2\rangle\langle |\varphi^*(k_3, t)|^2\rangle} . \tag{9.1}$$

With data from the poloidal probe array the bicoherence is calculated directly in k-space, which allows for the study of the nonlinear coupling with the zonal flow. Also, the Taylor hypothesis is not needed. A temporal average is used as ensemble average $\langle \cdot \rangle$, where, when conditioned on zonal potential fluctuations, the dynamics around the zonal flow occurrence are obtained (Sect. 9.1.1). The coupling strength is then scaled with collisionality (Sect. 9.1.2).

9.1.1 Bicoherence during zonal flow occurrence

To obtain the bicoherence spectrum corresponding to the spectral transfer of density fluctuation activity, $b_{nn\phi}^2$, density and potential fluctuations measured simultaneously on two neighbouring flux surfaces are used (cf. Sect. 5.2.3). With density fluctuations from FS 2 and potential fluctuations from FS 3, the fluctuating quantities in formula (9.1) are assigned to $\varphi(k_1) = n(k_1)$, $\varphi(k_2) = n(k_2)$, and $\varphi(k_3) = \phi(k_3)$. For the bicoherence connected to the fluid kinetic energy, $b_{\phi\phi\phi}^2$, only potential fluctuations on FS 3 are used, i.e. $\varphi(k_1) = \phi(k_1)$, $\varphi(k_2) = \phi(k_2)$, and $\varphi(k_3) = \phi(k_3)$. The data from a measurement in helium at low magnetic field ($\#10003$) is analysed throughout this chapter.

In general, the presence of wave-wave interactions is a signature of turbulence. As the zonal flow is an intermittent event and its duration covers only less than 7 % of the full time trace, the fraction of its three-wave coupling in the overall spectrum is small. With the conditional average, only realisations around the zonal flow occurrence are extracted and used for the ensemble average.

In figure 9.1 the non-redundant part of the bicoherence spectrum is shown as contour plot. Above is the integrated bicoherence

$$b^2(k_3) = \sum_{\substack{k_1, k_2 \\ k_3 = k_1 + k_2}} b^2(k_1, k_2)\, \delta_{k_1 + k_2, k_3} \ , \tag{9.2}$$

which represents the overall coupling with the k_3 potential mode. In the left figure (a) the axes k_1 and k_2 show wavenumbers of potential modes, whereas in the right figure (b) they stand for density modes. The zonal flow coupling is, therefore, shown on the $k_2 = -k_1$ line (counter diagonal) and, in the case of $b_{\phi\phi\phi}^2$, additionally on the horizontal line where $k_2 = 0$.

The main phase coherence in both spectra is mainly restricted to the zonal flow coupling.[1] For $b_{nn\phi}^2$ some mode coupling occurs at $k_2 \approx -12\,\mathrm{m}^{-1}$

---

[1]It has to be stressed that the bicoherence spectrum of the full signal is rich in mode coupling as three-wave interaction is an important mechanism in turbulence but, here, the data is filtered (conditional averaging) to obtain the drift-wave zonal-flow interaction.

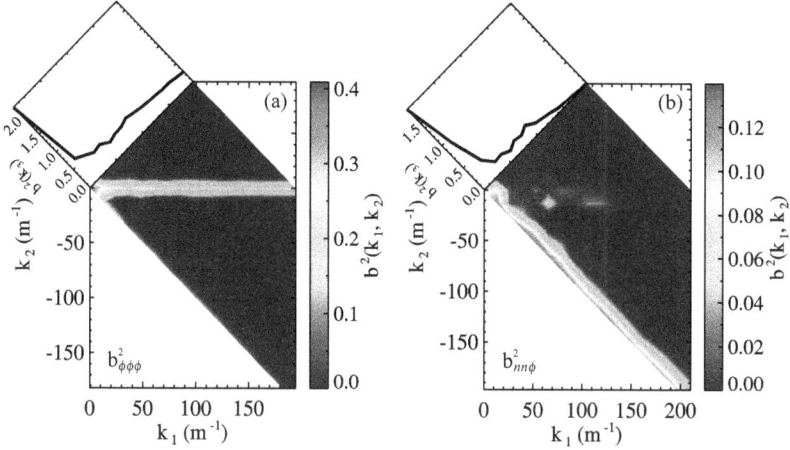

Figure 9.1: Bicoherence spectrum (contour plot) and integrated bicoherence (line plot) calculated as auto (a) and cross-spectrum (b) from density $n(k_1, k_2)$ and potential fluctuations $\phi(k_3)$. For the ensemble average subwindows of 256 µs width around the zonal flow occurrence are averaged over. The zonal flow is the $k_3 = 0$ component, which exhibits a broad spectrum.

with the maximum at $k_1 \approx 61\,\mathrm{m}^{-1}$ and $k_3 \approx 49\,\mathrm{m}^{-1}$. This is not directly connected to the zero potential mode and represents the coupling of the $m = 1$ density mode. In toroidal configurations the GAM is a candidate for this kind of density perturbation and the coincident appearance with the $m = 0$ potential mode does indeed suggest such an instability. A connection to the GAM and the geodesic energy transfer will be further discussed in section 9.2.1.

The bispectra show broad mode coupling along the zonal flow occurrence. However, this includes the coupling preceding and following the zonal flow as the ensemble average is calculated over the entire subwindows around the trigger time points. With the conditional averaging technique it is also possible to resolve the temporal evolution. This gives the dynamics of the nonlinear mode coupling with respect to the trigger time point, which, as in the analyses before, is chosen to be at the zonal potential maximum. A 2σ trigger condition is used with a subwindow size of 256 µs. The results for $b^2_{\phi\phi}$ and $b^2_{nn\phi}$ are shown in figure 9.2 on the left and right hand side, respectively. At the top (Fig. (a,b)), the total bicoherence is shown

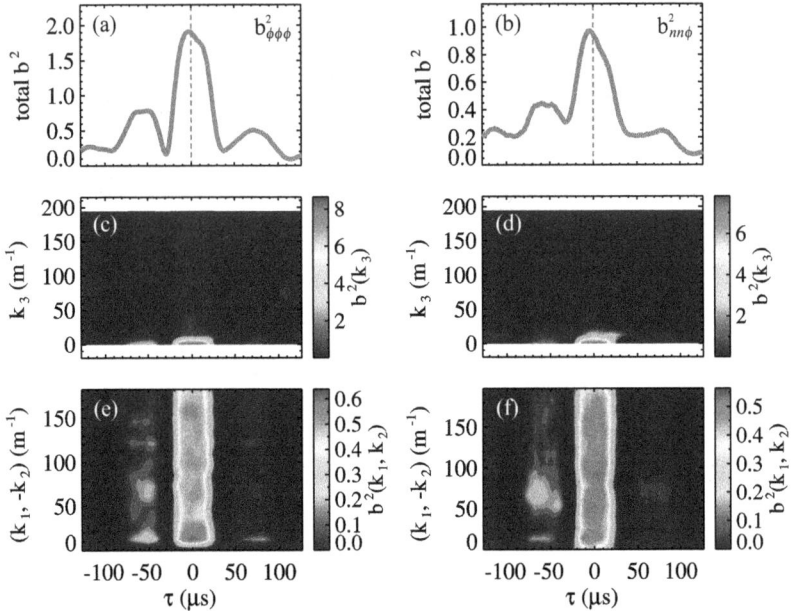

Figure 9.2: Time resolved nonlinear coupling around the maximum of the zonal potential ($\tau = 0\,\mu s$). Data of $b^2_{\phi\phi\phi}$ is given on the left and of $b^2_{nn\phi}$ on the right. At the top (a,b) the total bicoherence is shown, which is a measure for the overall nonlinear coupling in the spectrum. The integrated bicoherence (c,d) and the bicoherence of the modes which couple to the zonal flow, i.e. $k_1 + k_2 = 0$, (e,f) are presented below. An oscillatory behaviour of the coupling strength, which built up prior to the zonal flow occurrence, is apparent.

$$\text{total } b^2 = \sum_{k_1,k_2} b^2(k_1, k_2) , \qquad (9.3)$$

which is the sum over the whole spectrum and indicates general nonlinear mode coupling in the turbulence. For a better visualisation of the zonal flow coupling only special parts of the bispectrum are depicted. Figures (c,d) show the evolution of the integrated bicoherence spectrum (Eq. (9.2)), where at $k_3 = 0$ is the coupling with the zero potential mode. Below (Fig. (e,f)), the evolution of the bicoherence spectrum is shown for the same time range. Here, only modes which couple to the zonal flow are chosen, i.e. which fulfil the resonance condition $k_2 = -k_1$.[2] For both bispectra, $b^2_{\phi\phi\phi}$ (a,c,e) and $b^2_{nn\phi}$ (b,d,f), an oscillatory behaviour is found which builts up towards the trigger time point and gets maximal just before the maximum of the zonal potential. As the bicoherence is representative of the Reynolds stress (Chap. 3.2), this is an additional indication of the Reynolds stress drive (Chap. 8.2.2). The integrated bicoherence (c,d) shows that the mode coupling is concentrated at the $k=0$ component. When the coupling is resolved for the different modes which can interact with the zonal potential mode (e,f), it can be seen that the coupling is generally broad. Prior to the zonal flow maximum ($\tau \approx -70\,\mu s$) the coupling with a $m=4$ mode ($k_1 = -k_2 \approx 53\,m^{-1}$) is apparent, more prominent in $b^2_{nn\phi}$ (f), which might be identified with the dominant drift wave structure (see Chap. 6.3.2). However, various modes couple to the zonal flow where the contribution of the modes with mode numbers $m = 1, 4, 7$, and 12 are especially strong around $\tau = 0\,\mu s$. The coupling with the different modes is discussed in section 9.2.1 in more detail. As a next step, the collisionality scaling of the coupling strength is examined.

9.1.2 Collisional scaling of coupling strength

As the bicoherence is a measure of the nonlinear mode coupling strength, it is connected to the Reynolds stress and, therefore, to the drive of the zonal flow. Figure 9.3 shows the scaling of the integrated bicoherence $b^2(k_3)$ of the $k_3 = 0$ component, indicative of the overall zonal flow coupling. The bispectrum is calculated as in the section before where the average includes all time points in the subwindows around the trigger time points. Measurements at low and high magnetic field are shown separately in figures (a)

---

[2]For plot 9.2 (e) the ($k_1 = 0$, $k_2 = 0$, $k_3 = 0$) coupling is suppressed as it does naturally outrange the other couplings.

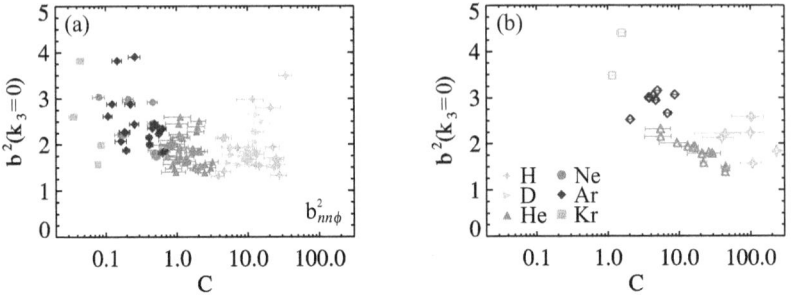

Figure 9.3: Scaling of the cross-bicoherence $b^2_{nn\phi}$ at $k_3 = 0$ (zonal potential mode) with collisionality for measurements at low (a) and high magnetic field (b). A higher bicoherence shows increased nonlinear coupling and indicates stronger zonal flow drive.

and (b), respectively. Especially for high magnetic field a clear increase in coupling strength is found with lower collisionality. The trend for low magnetic field (Fig. (a)) is similar but not as clear, and values for deuterium are, again, relatively high when compared to those of hydrogen. This is indicative of an increased Reynolds stress drive with lower collisionality which would explain the increase in zonal flow power for stronger adiabatic coupling (cf. Sect. 7.3.2). However, the bicoherence includes all mode couplings and does not distinguish between drive and damping. In order to obtain only the drive of the zonal flow also the nonlinear coupling coefficients have to be considered. Therefore, the wave kinetic equation has to be solved in order to obtain nonlinear power transfer function, which is the topic of the next section.

9.2 Analysis of energy transfer channels

As stated by the wave kinetic equation (4.35), the evolution of the energy can be separated into a linear and nonlinear part,

$$\frac{\partial}{\partial t} E_{k_3} = \text{linear terms} + \sum_{k_3 = k_1 + k_2} T(k_3 \leftrightarrow (k_1, k_2)) , \tag{9.4}$$

where the amount and direction of the spectral power transfer connected to three-wave interactions is given by the nonlinear energy transfer function

$$T(k_1, k_2) = \mathrm{Re}(\Lambda_{k_3}^Q(k_1, k_2)\langle\varphi(k_1, t)\varphi(k_2, t)\varphi^*(k_3, t)\rangle) \,. \tag{9.5}$$

This includes the bispectrum $\langle\varphi(k_1, t)\varphi(k_2, t)\varphi^*(k_3, t)\rangle$ and the coupling coefficients $\Lambda_{k_3}^Q(k_1, k_2)$, which can be estimated from experimental data as described in chapter 4.5. For the present analysis of the energy transfer with the zonal flow the modified Ritz method proposed by Kim et al. [153] is used to obtain the energy transfer function. Again, the poloidal probe array permits the calculation of equation (9.5) directly in wavenumber space with the assignment of the fluctuating quantities stated in section 9.1.1, i.e.

$$T^{\mathrm{V}}(k_1, k_2) = \mathrm{Re}(\Lambda_k^Q(k_1, k_2)\langle\phi(k_1)\phi(k_2)\phi^*(k_3)\rangle) \,, \tag{9.6}$$

$$T^{\mathrm{N}}(k_1, k_2) = \mathrm{Re}(\Lambda_k^Q(k_1, k_2)\langle n(k_1)n(k_2)\phi^*(k_3)\rangle) \,. \tag{9.7}$$

Instead of the full 2D treatment of the turbulence only the 1D-part, covering poloidal mode coupling, is used (following [219]) which is most relevant for the zonal flow interaction. As for the bispectral analysis (cf. Sect. 9.1), the conditional average is applied in equation (9.5) to obtain the dynamics around the zonal flow occurrence (Sect. 9.2.1). The resulting transfer rates of the different coupling channels are then analysed for their collisional scaling (Sect. 9.2.2).

9.2.1 Energy transfer with the zonal flow

The Kim method also considers fourth-order moments in order to avoid a closure approximation (see Sect. 4.5.2). In contrast to the Ritz method it is thus more robust against the influence of noise and fluctuations not covered by three-wave interactions.[3] Involving the next higher-order moment in the calculation of the coupling coefficients makes the method more applicable but results in greater computational cost and induces an potential error source since the calculation involves matrix inversions. However, this method is best suited for the present analysis as a good k-space resolution simultaneously for density and potential modes can be retained.[4]

---

[3]The Kim method in its original form separates ideal and non-ideal fluctuations to validate the method. This distinction will not be made for the experimental data and the whole signal is treated as ideal fluctuations, which satisfy the wave-coupling equation (4.28).

[4]In case of the Camargo method [39] the turbulence has to be treated intrinsically two-dimensional. For the free-energy-like transfer function T^{N} the maximal resolvable mode number would be reduced by a factor of 2 [148].

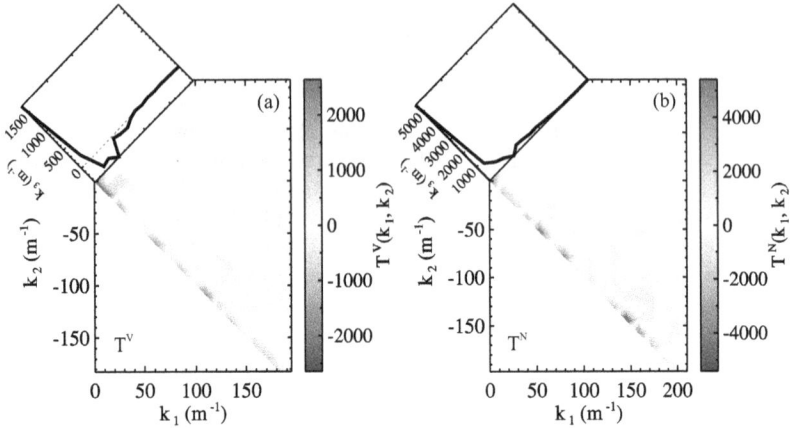

Figure 9.4: Energy transfer function averaged around the zonal flow occurrence. The line plot gives the energy transfer with the resonant mode $k_3 = k_1 + k_2$, where positive values indicate energy gain and negative show energy loss. Various modes, up to smallest resolvable scale, contribute to the drive of the zonal flow ($k_3 = 0$). In figure (a) the energy transfer of the potential fluctuations and in (b) the energy transfer of density fluctuations is shown.

Following the procedure in section 9.1, the energy transfer functions are calculated around the zonal flow occurrence. Figure 9.4 (a) shows the kinetic-energy-like transfer function T^V and (b) the free-energy-like transfer function T^N. Equivalent to the integrated bicoherence, the integrated transfer function, $T(k_3) = \sum_{k_1,k_2} T_k(k_1, k_2) \delta_{k_1+k_2,k_3}$, is included in the figure. The interpretation of the different axes is the same as in figure 9.1, where the zonal flow is represented as the $k_3 = 0$ component. The transfer function $T(k_1, k_2)$ will attain positive values when mode k gains energy ($(k_1, k_2) \rightarrow k$), and negative values when energy in mode k is transferred to modes k_1 and k_2 ($(k_1, k_2) \leftarrow k$). Due to the chosen average, the transfer rates connected with the zonal potential mode dominate the spectrum. For the zonal flow ($m = 0$) a positive value for the energy transfer is found, confirming the inverse energy transfer originally published in [34] obtained by the Camargo method [39]. Various modes, especially in the mid- and higher wavenumber range ($k_1 = -k_2 > 70\,\mathrm{m}^{-1}$), contribute positively in the energy transfer with the zonal flow. This points to a non-local interaction with the zonal flow where smaller structures transfer their energy directly to the zero mode. A possible energy transfer to the GAM ($m=1$) might be inferred from the negative transfer rates at ($k_1 = -k_2 = 12\,\mathrm{m}^{-1}$). However, the transfer function is calculated as an average over all time points around the zonal flow occurrence adding up positive and negative contributions.

With the conditional averaging procedure the temporal evolution around the zonal flow occurrence can be analysed. The representation of the different parts of the transfer function is the same as in figure 9.2. Similar to the total bicoherence the total energy transfer may be defined, total $T = \sum_{k_1,k_2} T(k_1, k_2)$, which is shown in figures 9.5 (a,b). Figures (c,d) show the integrated power transfer, and in figure (e,f) the transfer rates for all modes which couple to the zero potential mode are depicted.[5]

The evolution of both transfer functions, T^V and T^N, is similar. Prior to the zonal flow maximum ($\tau = 0\,\mathrm{\mu s}$) the overall energy transfer is positive, and it gets negative as the zonal flow decays. The integrated bicoherence in figure (c,d) confirms that this energy transfer is mainly restricted to the zonal flow ($k_3 = 0$). Broken down into the individual mode coupling ($k_1 = -k_2$ line) reveals a complex transfer pattern along the zonal flow evolution (Figs. (e,f)). As for the bicoherence, the oscillatory behaviour is observed (mainly in T^V), which culminates in a broad energy transfer shortly before the zonal flow maximum. Modes up to the highest resolvable wavenumbers transfer energy to the zonal flow, but also a contribution at $k_1 = -k_2 = 12\,\mathrm{m}^{-1}$ is visible, which could be related to the GAM as it possesses mode number $m = 1$. While the energy input from large scale structures ($k_1 = -k_2 < 66\,\mathrm{m}^{-1}$) quickly ceases away, becoming a sink for the zonal flow energy, selected small scale (density) structures ($k_1 = -k_2 > 66\,\mathrm{m}^{-1}$) keep on pumping energy to the zonal flow although the flow maximum is exceeded (Fig. (f)). This corresponds to the findings in chapter 8.2.2 where, especially for small scales, the influence on the phase is large for the full flow existence. When the zonal flow decays distinct modes occur, which gain energy from the zonal flow. For the density the transfer to the $m=6$ mode ($k_1 = -k_2 = 79\,\mathrm{m}^{-1}$) is especially strong. This shows that, besides other damping mechanisms, nonlinear mode coupling plays an important role for the zonal flow damping. Naively, such a mode could be thought of as the GAM in stellarator geometry. However, the geodesic transfer effect (Chap. 3.4.2) is, at first, a linear

---

[5]The current analysis does not show the transfer of kinetic and free-energy in the overall turbulence but is restricted to the drift-wave zonal-flow interaction by applying the conditional average. Thus, the usual direction of the energy transfer must not be expected.

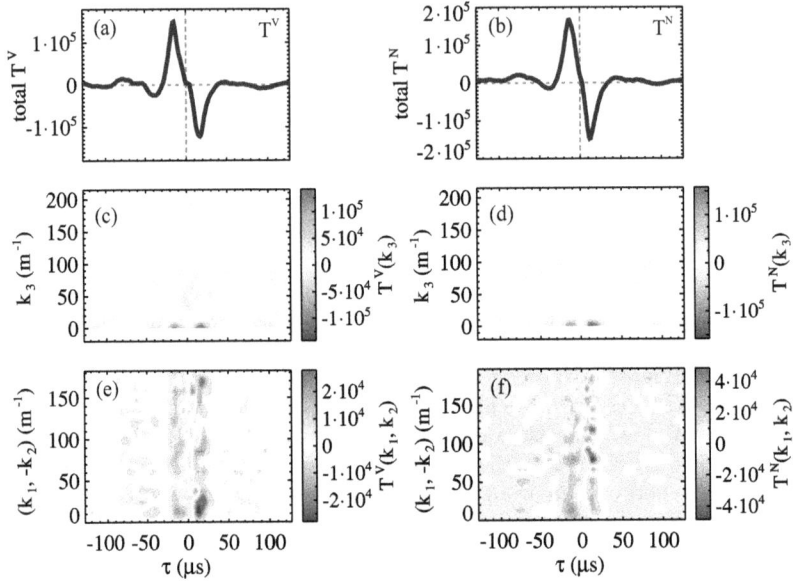

Figure 9.5: Temporal evolution of the nonlinear energy transfer function T^V (left) and T^N (right) around the zonal flow occurrence. In figure (a,b) the total energy transfer, in (c,d) the integrated energy transfer, and in (e,f) the energy transfer with the zonal flow is shown (cf. Fig. 9.2). The nonlinear energy transfer is important for the zonal flow growth ($\tau \leq 0$) and its damping ($\tau > 0$). Especially, the energy transfer to the $k_1 = -k_2 = 79\,\mathrm{m}^{-1}$ density mode ($m=6$) is high.

damping mechanism, and the GAM structure in non-axial magnetic fields is complicated as several higher Fourier components of the magnetic field have to be considered. For the magnetic field in TJ-K the $m = 1$ and $m = 6$ components are indeed the dominating contributions [220].[6] The results in figure 9.5, thus, suggest a contribution to the zonal potential attribute to the GAM, which cannot be separated from the zonal flow in the current analysis.

Nevertheless, positive and negative energy transfer rates of the $k_3 = 0$ component can now be distinguished. This is used to analyse the collisionality scaling of the different transfer channels as a next step.

9.2.2 Influence of collisionality

With the wavenumber resolved energy transfer function in combination with the conditional ensemble average it is possible to examine the collisionality scaling of energy gain and loss of the zonal flow. To obtain a representative value for each transfer channel, the time resolved transfer function is averaged over a certain time window asymmetrically positioned with respect to the trigger time point (zonal flow maximum). As the cross-coupling between density and potential is of interest, the free-energy-like transfer function T^N is used for the analysis.

In figure 9.6 the scaling of the transfer rates of specific modes is shown for measurements at low (left) and high magnetic field (right) separately as the magnitude in some instances differs significantly. For the scaling of the energy transfer into the $m = 0$ mode, i.e. zonal flow drive, the energy transfer function is averaged over $128\,\mu s$ prior to the zonal flow maximum. Compared with figure 9.5 this includes most of the positive contributions of the transfer function. Figures (a) and (b) confirm the trend observed in the scaling of the bicoherence (Sect. 9.1.2) as for decreasing collisionality the energy transfer to the zonal flow clearly increases. For high magnetic field (a) the increase is especially strong and a fit with a power law C^α yields a value of $\alpha = -0.60 \pm 0.07$. Noticeable are, again, the high values for deuterium in comparison with hydrogen (cf. Chap. 7.3.2). Higher energy transfer rates might explain the increased zonal flow level in deuterium but the reason for this isotope effect remains unclear as the cross-field coupling is not significantly higher (cf. Chap. 8.2.3). For low magnetic field the power

---

[6]For the actual GAM structure also the damping at different mode numbers has to be considered, which increases with e^{-1/n^2} [108].

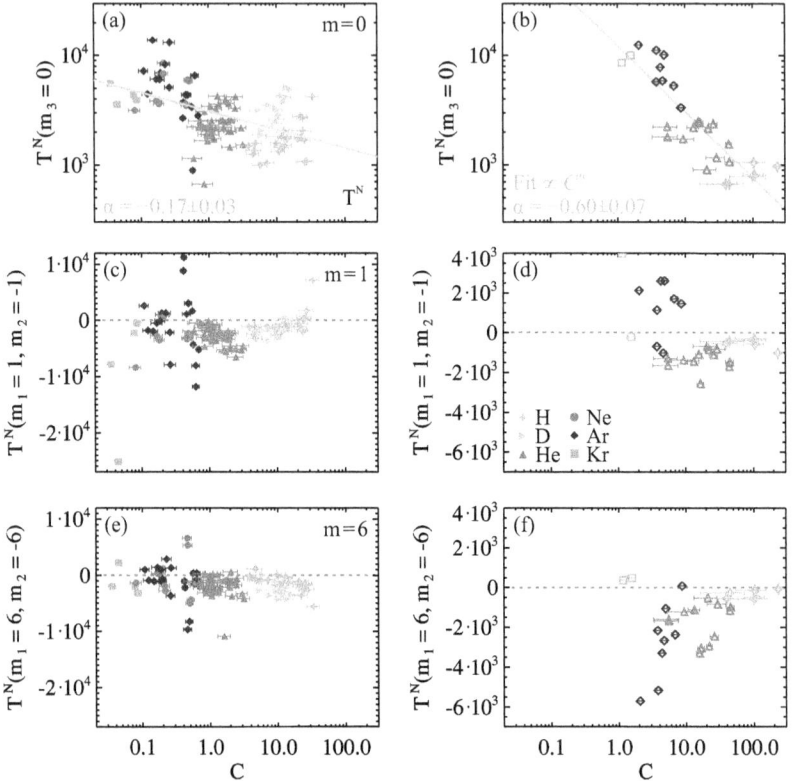

Figure 9.6: Collisional scaling of the nonlinear energy transfer for specific modes. Measurements at low magnetic field are shown on the left and at high magnetic field on the right. The energy transfer to the zonal flow is captured by the transfer function at $k_3 = 0$ ($m = 0$) averaged over $128\,\mu s$ before the zonal flow maximum (a,b). The results of a power law fit are given in the figure. For the $m = 1$ (c,d) and $m = 6$ (e,f) density modes the transfer function is mostly negative when averaged over the time after the zonal flow maximum ($32\,\mu s$). Thus, these modes are loss channels for the energy in the zonal potential mode but, in both cases, no clear scaling with collisionality is found.

law fit results in $\alpha = -0.17 \pm 0.03$ when all gases are considered and $\alpha = -0.23 \pm 0.03$ without deuterium. These are similar values as found for the collisional scaling of the zonal flow power in chapter 7.3.2 which underlines the close connection of both quantities.

The analysis in the previous section showed that a negative energy transfer occurs as the zonal flow decays. This motivates the study of different transfer channels which could act as loss channels for the energy in the zero potential mode. For the GAM in the magnetic field structure of TJ-K the $m = 1$ component (tokamak mode) and the $m = 6$ component (due to the sixfold symmetry of the experiment) are expected. Therefore, the respective energy transfer of the zero potential mode to these two components is shown in figure 9.6 (c-f). The transfer function is averaged over $32\,\mu s$ following the maximum of the zonal potential. For the $m = 1$ (c,d) and $m = 6$ density mode (e,f) mostly negative values are found for the energy transfer, which shows that both components are loss channels for the zonal potential mode. The collisionality scaling is, however, inconclusive. For measurements at low magnetic field the $m = 1$ transfer channel (Fig. 9.6 (c)) shows a tendency to increased values with lower collisionality (higher zonal flow power). For high magnetic field the $m = 6$ transfer channel (Fig. 9.6 (f)) shows this trend. As the geodesic coupling is complex in stellarators, the role of the different modes cannot be fully resolved.

9.3 Summary of the chapter

As a modulation instability, the zonal flow is driven nonlinearly by the turbulent plasma modes. The bicoherence, as a measure for the nonlinear coupling strength, and the nonlinear energy transfer function are calculated using experimental data. Performing the calculation in k-space together with a conditional average allows for the direct analysis of the zonal flow interaction. The results can be summarised as follows:

- The bispectral analysis, indicative of three-wave interaction, reveals that with the zonal flow occurrence the nonlinear coupling is strong, which demonstrates the Reynolds stress drive of the zonal flow in another way. The coupling is generally broad where high values of phase coherence are found up to the smallest resolvable scale. An oscillatory behaviour in the coupling strength is observed, but also the coupling to the $m = 1$ density mode is apparent which suggests the presents of a GAM oscillation.

- With the calculation of the energy transfer function the direction and the amount of energy transfer is resolved. Prior to the zonal potential maximum the energy transfer is strongly positive confirming the picture of the nonlinearly driven zonal flow. Furthermore, energy transfer from small scale structures support the picture of a nonlocal driving mechanism. Also for the decaying zonal flow the nonlinear energy transfer is high, where, especially, the transfer to the $m = 6$ density mode is prominent. This could be connected to the GAM, which exhibits the sixfold symmetry of the magnetic field.

- The collisional scaling of the bicoherence shows an increased coupling with the zonal flow for lower collisionality, indicating increased drive. Similarly, the energy transfer to the zonal flow increases for lower collisionality, when the transfer function is averaged prior to the zonal flow maximum. As for the relative zonal flow power, deuterium shows higher values of the transfer rates despite of a similar collisionality as hydrogen. The $m = 1$ and $m = 6$ density modes (GAM) appear as loss channels but a clear scaling with collisionality is not observed.

Chapter 10
Summary and conclusion

Turbulence in two dimensions tends to self-generate large-scale turbulent structures, so-called zonal flows. These shear flows play an important role in fusion research as they are thought to be connected to the transition to a high confinement regime (H-mode). However, the physics of zonal flows in plasmas, especially in complex magnetic field geometries, is not sufficiently understood in theory and experiment. Primarily, this is due to the fact that a realistic treatment of the magnetic field in simulations is difficult and measurements of turbulent fluctuations are hard to obtain in fusion plasmas. The measurements for this work have been conducted at the stellarator experiment TJ-K. Although the plasma parameters are comparatively low, it has been shown that normalised quantities are similar to those in fusion edge plasmas. The low temperatures allow the use of Langmuir probes in the entire confinement region. With a poloidal probe array, consisting of 128 Langmuir probes with 32 probes on each of four neighbouring magnetic flux surfaces, density and potential fluctuations have been acquired in this work with high spatial and temporal resolution at the same time. Thus, velocity fluctuations as well as the turbulent Reynolds stress, and its gradient, are available on the complete poloidal circumference. This gives the unique possibility to directly study the zonal flow and the connected turbulent dynamics in a toroidally confined plasma. This work concentrates on the investigation of the Reynolds stress drive of zonal flows with its connection to the geometry of the confining magnetic field. The special focus is on its collisionality dependence, which determines the cross-coupling between the density and potential fluctuations. In order to gradually change the collisionality, a multitude of measurements has been performed, using the gases H_2, D_2, He, Ne, Ar, and Kr at different pressure and heating power.

The main results of this work can be summarised as follows:
All measurements of this work, performed using newly designed poloidal limiters, exhibit centrally peaked density profiles and electron temperature profiles with a maximum in the edge of the plasma. Mainly by changing

the ion mass, the collisionality C could be varied by about four orders of magnitude, which makes it possible to study the transition from the hydrodynamic regime $(C \gg 1)$ to the adiabatic regime $(C \ll 1)$. As found in simulations, the density and potential fluctuation levels are found to scale with collisionality, pointing to a destabilisation of the drift waves through an altered density-potential coupling. The spectra show the typical shape of turbulence and scale with the structure size, i.e $\rho_s^{0.43}$.

In the wavenumber frequency spectra a $k_\theta = 0$ mode in the potential is apparent, while not present in the density. This is the signature of the zonal flow, which is known to be a pure potential mode, and it also excludes the possibility of a pure mean background fluctuation since the density is not changed. Positive and negative zonal potential fluctuations appear equally distributed in a burst like, intermittent manner, and the frequency spectra show that the main spectral power is located below 8 kHz. Measurements with the poloidal probe array in combination with the 2D-movable probe unit clearly show the 3D structure of the zonal flow as a homogeneous potential perturbation on the whole flux surface narrow in its radial extent. With $k_r \approx 50\,\mathrm{m}^{-1}$ the radial wavenumber is in the range of the dominant drift wave structures.

The tilt of vortices in a shear flow is the fundamental principle of the Reynolds stress drive of zonal flows. It is found that the time-averaged (mean) Reynolds stress is not homogeneously distributed along the flux surface but has a strong poloidal asymmetry where it is concentrated in regions with negative normal magnetic curvature κ_n and positive geodesic curvature κ_g. This is similar to the distribution of the turbulent cross-field transport, which is plausible by reason of similar conceptual form of both quantities. Also, integrated magnetic shear as well as local magnetic shear have an influence on the tilt of the turbulent structures even though mostly overlaid by the curvature effects. Asymmetries in the Reynolds stress profile have been confirmed by simulations, which substantiates the nonlocality of the background shear flow formation.

Using a conditional averaging technique, the evolution in a time window around the zonal flow occurrence was analysed. A direct comparison of the driving term and the poloidal flow illustrates the Reynolds stress drive of the zonal flow, with a spatial structure similar to that of the mean Reynolds stress, suggesting an analogue influence of the background magnetic field. Also in higher-order spectra the drift-wave zonal-flow interaction is apparent. The nonlinear coupling, measured by the bicoherence, is strong around the zonal flow occurrence and the energy transfer from small scale structures supports the picture of a nonlocal driving mechanism (k-space). With the possibility of Reynolds stress measurements in real space, it was found that for increasing collisionality the coupling between density and potential decreases, which in return makes the zonal flow driving mechanism less effective. As a result, also the nonlinear energy transfer into the zonal flow, as well as the relative spectral power of the zonal flow, decreases with higher collisionality. A power law fit C^α yields a value of $\alpha = -0.19 \pm 0.02$ for the zonal flow power where values of up to 29 % of the total turbulent spectral power are reached. This is a direct test of a fundamental mechanism in plasma turbulence on a microscopic level of the fluctuations and represents a first verification of the importance of collisionality for large-scale structure formation in magnetically confined toroidal plasmas. In the scaling analyses, an isotope effect is observed for deuterium where the zonal flow power and the energy transfer rates are higher as compared to hydrogen although the discharges have comparable collisionalities. The cause for this increase could not be identified in the present experiments.

An additional peak in the frequency spectra of the zonal potential at higher frequencies suggests the presents of a geodesic acoustic mode (GAM). The analysis of the bicoherence spectrum and the energy transfer function, where an interaction with the $m = 1$ and $m = 6$ modes appears, supports this conclusion. However, the density structure of the GAM in non-axial magnetic fields is complicated and, with the present analysis methods, difficult to separate from the low frequency zonal flow fluctuation.

Outlook

Some aspects of zonal flows could be analysed in this work but many questions have to be left unanswered. The turbulence in the adiabatic regime (low collisionality) exhibits a strong zonal potential contribution with an additional fluctuation at higher frequencies. It has to be clarified if this high frequency component of the zonal potential is indeed a GAM oscillation. On the one hand this requires the calculation of the expected density structure in the magnetic field geometry of the experiment and on the other hand a scaling analysis of the natural frequency with the sound speed. The zonal flow can couple to the GAM, and the presence of both modes could result in interesting three species dynamics. Both modes would have to be separated which might be accomplished by using data mining algorithms, viz. clustering analyses, on the zonal potential.[1] This can also be used to detect different scenarios of the zonal flow dynamics and, thus, to improve the conditional averaging technique.

In here, the zonal flow has been analysed in the poloidal cross section but the flow is inherently three-dimensional. With the combination of the poloidal probe array and the 2D-movable probe, e.g. with Mach probes, also the parallel flow can be studied. This could show the $m = 1$ asymmetry of the flow, originating from toroidicity, and possible higher modes which would give a hint on the actual GAM structure. With regard to the Reynolds stress drive of zonal flows, the role of the parallel component of the Reynolds stress tensor can be analysed.

Concerning the magnetic field dependence of the Reynolds stress, the influence of the different field parameters has to be unravelled. This is a complicated task as the influences cannot completely be separated, and further analyses require expensive measurements at different toroidal positions. As all other investigations, such studies should be accompanied by simulation which would help with the interpretation of the data.

The measurements with deuterium show an effect of the isotope mass on the zonal flow activity. Further measurements, also at high magnetic field, are needed to verify this trend and to resolve the cause of this isotope effect.

Background shear flows are linked to turbulent shear flows, like the zonal flow or GAM, via manifold shrinking, which reduces the turbulent mode coupling to the zonal flow interaction. Using plasma biasing, shear flows can be effectively induced in the experiment. The linkage between background and turbulent flow is an interesting phenomena, and, due to the flow enhancement, it might give the possibility to access further dynamical states of the turbulent system.

---

[1]Since the number of clusters is not known in advance, density-based hierarchical clustering algorithms seem practicable (see OPTICS algorithm [221, 222]).

Bibliography

[1] J. D. Lawson, Proceedings of the Physical Society. Section B **70**, 6 (1957).

[2] R. J. Goldston, Plasma Physics and Controlled Fusion **26**, 87 (1984).

[3] F. Wagner, The European Physical Journal H **43**, 523 (2018).

[4] F. Wagner, Tech. Rep., Max-Planck-Institut für Plasmaphysik, Garching (2017).

[5] F. Wagner and U. Stroth, Plasma Physics and Controlled Fusion **34**, 1803 (1992).

[6] F. Wagner *et al.*, Physical Review Letters **49**, 1408 (1982).

[7] ASDEX Team, Nuclear Fusion **29**, 1959 (1989).

[8] F. Wagner, Plasma Physics and Controlled Fusion **49**, B1 (2007).

[9] F. Wagner *et al.*, Physical Review Letters **53**, 1453 (1984).

[10] H. Biglari, P. H. Diamond, and P. W. Terry, Physics of Fluids B: Plasma Physics **2**, 1 (1990).

[11] C. P. Ritz, H. Lin, T. L. Rhodes, and A. J. Wootton, Physical Review Letters **65**, 2543 (1990).

[12] K. H. Burrell, Physics of Plasmas **6**, 4418 (1999).

[13] P. W. Terry, Reviews of Modern Physics **72**, 109 (2000).

[14] B. A. Carreras, V. E. Lynch, and L. Garcia, Physics of Fluids B: Plasma Physics **3**, 1438 (1991).

[15] P. H. Diamond and Y.-B. Kim, Physics of Fluids B: Plasma Physics **3**, 1626 (1991).

[16] E.-j. Kim and P. H. Diamond, Physical Review Letters **90**, 185006 (2003).

[17] P. Manz *et al.*, Physics of Plasmas **19**, 072311 (2012).

[18] K. Miki, P. H. Diamond, S. H. Hahn, W. W. Xiao, Ö. D. Gürcan, and G. R. Tynan, Physical Review Letters **110**, 195002 (2013).

[19] Y. H. Xu, C. X. Yu, J. R. Luo, J. S. Mao, B. H. Liu, J. G. Li, B. N. Wan, and Y. X. Wan, Physical Review Letters **84**, 3867 (2000).

[20] R. A. Moyer, G. R. Tynan, C. Holland, and M. J. Burin, Physical Review Letters **87**, 135001 (2001).

[21] T. Estrada, T. Happel, C. Hidalgo, E. Ascasíbar, and E. Blanco, Europhysics Letters **92**, 35001 (2010).

[22] G. D. Conway, C. Angioni, F. Ryter, P. Sauter, and J. Vicente, Physical Review Letters **106**, 065001 (2011).

[23] L. Schmitz *et al.*, Physical Review Letters **108**, 155002 (2012).

[24] G. Tynan *et al.*, Nuclear Fusion **53**, 073053 (2013).

[25] Z. Yan, G. R. McKee, R. Fonck, P. Gohil, R. J. Groebner, and T. H. Osborne, Physical Review Letters **112**, 125002 (2014).

[26] T. Kobayashi *et al.*, Physical Review Letters **111**, 035002 (2013).

[27] J. Cheng *et al.*, Nuclear Fusion **54**, 114004 (2014).

[28] M. Cavedon *et al.*, Nuclear Fusion **57**, 014002 (2017).

[29] F. H. Busse, Pure and Applied Geophysics **121**, 375 (1983).

[30] P. B. Rhines, Chaos: An Interdisciplinary Journal of Nonlinear Science 4, 313 (1994).

[31] H. Branover, A. Eidelman, E. Golbraikh, and S. Moiseev, *Turbulence and Structures: Chaos, Fluctuations, and Helical Self-organization in Nature and the Laboratory* (Academic Press, San Diego, 1999).

[32] C. Brandt, IPP Greifswald.

[33] P. H. Diamond, S.-I. Itoh, K. Itoh, and T. S. Hahm, Plasma Physics and Controlled Fusion **47**, R35 (2005).

[34] P. Manz, M. Ramisch, and U. Stroth, Physical Review Letters **103**, 165004 (2009).

[35] P. Manz, G. Birkenmeier, M. Ramisch, and U. Stroth, Physics of Plasmas **19**, 082318 (2012).

[36] P. Manz, M. Ramisch, and U. Stroth, Physical Review E **82**, 056403 (2010).

[37] U. Stroth, P. Manz, and M. Ramisch, Plasma Physics and Controlled Fusion **53**, 024006 (2011).

[38] G. Birkenmeier, M. Ramisch, B. Schmid, and U. Stroth, Physical Review Letters **110**, 145004 (2013).

[39] S. J. Camargo, D. Biskamp, and B. D. Scott, Physics of Plasmas **2**, 48 (1995).

[40] A. Fujisawa et al., Nuclear Fusion **47**, S718 (2007).

[41] A. Fujisawa, Nuclear Fusion **49**, 013001 (2009).

[42] C. L. M. H. Navier, Mémoires de l'Académie Royale des Sciences de l'Institut de France **6**, 389 (1823).

[43] G. G. Stokes, Transactions of the Cambridge Philosophical Society **8**, 287 (1845).

[44] H. Rose and P. Sulem, Journal de Physique **39**, 441 (1978).

[45] G. K. Batchelor, *An Introduction to Fluid Mechanics* (Cambridge University Press, Cambridge, 1967).

[46] P. Manneville, *Dissipative Structures and Weak Turbulence* (Academic Press, San Diego, 1990).

[47] T. Klinger, A. Latten, A. Piel, G. Bonhomme, T. Pierre, and T. Dudok de Wit, Physical Review Letters **79**, 3913 (1997).

[48] T. Klinger, A. Latten, A. Piel, G. Bonhomme, and T. Pierre, Plasma Physics and Controlled Fusion **39**, B145 (1997).

[49] T. Klinger, C. Schröder, D. Block, F. Greiner, A. Piel, G. Bonhomme, and V. Naulin, Physics of Plasmas **8**, 1961 (2001).

[50] H. Lamb, *Hydrodynamics* (Dover publications, New York, 1945).

[51] S. Douady, Y. Couder, and M. E. Brachet, Physical Review Letters **67**, 983 (1991).

[52] D. Bonn, Y. Couder, P. H. J. van Dam, and S. Douady, Physical Review E **47**, R28 (1993).

[53] G. I. Taylor, Proceedings of the Royal Society of London. Series A - Mathematical and Physical Sciences **151**, 421 (1935).

[54] G. I. Taylor, Proceedings of the Royal Society of London. Series A - Mathematical and Physical Sciences **164**, 476 (1938).

[55] P. R. Halmos, *Lectures on ergodic theory* (AMS Chelsea Publishing, New York, 1956).

[56] E. Noether, Transport Theory and Statistical Physics **1**, 186 (1971).

[57] M. Steenbeck, F. Krause, and K.-H. Rädler, Zeitschrift für Naturforschung A **21**, 369 (1966).

[58] S. Gama and U. Frisch, in *Solar and Planetary Dynamos*, edited by M. R. E. Proctor, P. C. Matthews, and A. M. Rucklidge (Cambridge University Press, Cambridge, 2008), Publications of the Newton Institute, pp. 115–120.

[59] J.-J. Moreau, Comptes rendus de l'Académie des sciences Paris **252**, 2810 (1961).

[60] U. Stroth, *Plasmaphysik* (Vieweg+Teubner, Wiesbaden, 2011).

[61] Y. Gagne, E. J. Hopfinger, and U. Frisch, in *New Trends in Nonlinear Dynamics and Pattern-Forming Phenomena*, edited by P. Coullet and P. Huerre (Springer, New York, 1990), pp. 315–319.

[62] K. R. Sreenivasan, Physics of Fluids **27**, 1048 (1984).

[63] A. N. Kolmogorov, Doklady Akademiia Nauk SSSR **32**, 16 (1941).

[64] G. L. Eyink and K. R. Sreenivasan, Reviews of Modern Physics **78**, 87 (2006).

[65] A. N. Kolmogorov, Doklady Akademiia Nauk SSSR **30**, 301 (1941).

[66] R. H. Kraichnan, Physics of Fluids **10**, 1417 (1967).

[67] G. L. Eyink, Physica D: Nonlinear Phenomena **91**, 97 (1996).

[68] N. Mahdizadeh, F. Greiner, M. Ramisch, U. Stroth, W. Guttenfelder, C. Lechte, and K. Rahbarnia, Plasma Physics and Controlled Fusion **47**, 569 (2005).

[69] G. M. Jenkins and D. G. Watts, *Spectral analysis and its applications* (Holden-Day, San Francisco, 1968).

[70] H. L. Pécseli, *Fluctuations in Physical Systems* (Cambridge University Press, Cambridge, 2000).

[71] E. J. Powers, Nuclear Fusion **14**, 749 (1974).

[72] B. A. Carreras *et al.*, Physics of Plasmas **3**, 2664 (1996).

[73] S. Krasheninnikov, Physics Letters A **283**, 368 (2001).

[74] D. A. D'Ippolito, J. R. Myra, and S. I. Krasheninnikov, Physics of Plasmas **9**, 222 (2002).

[75] D. A. D'Ippolito, J. R. Myra, and S. J. Zweben, Physics of Plasmas **18**, 060501 (2011).

[76] B. Nold, G. D. Conway, T. Happel, H. W. Müller, M. Ramisch, V. Rohde, and U. Stroth, Plasma Physics and Controlled Fusion **52**, 065005 (2010).

[77] G. Fuchert, G. Birkenmeier, B. Nold, M. Ramisch, and U. Stroth, Plasma Physics and Controlled Fusion **55**, 125002 (2013).

[78] B. Nold *et al.*, Physics of Plasmas **21**, 102304 (2014).

[79] G. S. Lee and P. H. Diamond, Physics of Fluids **29**, 3291 (1986).

[80] F. Romanelli, Physics of Fluids B: Plasma Physics **1**, 1018 (1989).

[81] F. Jenko, W. Dorland, M. Kotschenreuther, and B. N. Rogers, Physics of Plasmas **7**, 1904 (2000).

[82] B. Labit and M. Ottaviani, Physics of Plasmas **10**, 126 (2003).

[83] T. Dannert and F. Jenko, Physics of Plasmas **12**, 072309 (2005).

[84] M. A. Leontovich, *Reviews of Plasma Physics* (Springer US, Boston, 1995).

[85] F. F. Chen, Physics of Fluids **8**, 912 (1965).

[86] B. Scott, Plasma Physics and Controlled Fusion **39**, 471 (1997).

[87] B. Scott, Plasma Physics and Controlled Fusion **39**, 1635 (1997).

[88] S. Niedner, B. D. Scott, and U. Stroth, Plasma Physics and Controlled Fusion **44**, 397 (2002).

[89] N. Mahdizadeh, F. Greiner, T. Happel, A. Kendl, M. Ramisch, B. D. Scott, and U. Stroth, Plasma Physics and Controlled Fusion **49**, 1005 (2007).

[90] K. Rahbarnia, E. Holzhauer, N. Mahdizadeh, M. Ramisch, and U. Stroth, Plasma Physics and Controlled Fusion **50**, 085008 (2008).

[91] G. Birkenmeier, M. Ramisch, G. Fuchert, A. Köhn, B. Nold, and U. Stroth, Plasma Physics and Controlled Fusion **55**, 015003 (2013).

[92] A. Hasegawa and M. Wakatani, Physical Review Letters **50**, 682 (1983).

[93] M. Wakatani and A. Hasegawa, Physics of Fluids **27**, 611 (1984).

[94] A. Hasegawa and K. Mima, Physical Review Letters **39**, 205 (1977).

[95] J. G. Charney, Geofysiske Publikasjoner **17**, 1 (1948).

[96] A. Hasegawa, C. G. Maclennan, and Y. Kodama, Physics of Fluids **22**, 2122 (1979).

[97] V. Naulin, *Physics of turbulent plasma*, Lecture notes (Christian-Albrechts-Universität zu Kiel, 2002).

[98] J. M. Dewhurst, B. Hnat, and R. O. Dendy, Physics of Plasmas **16**, 072306 (2009).

[99] J. M. Dewhurst, *Statistical Description and Modelling of Fusion Plasma Edge Turbulence*, Ph.d. thesis (University of Warwick, 2010).

[100] W. D. D'haeseleer, W. N. G. Hitchon, J. D. Callen, and J. L. Shohet, *Flux Coordinates and Magnetic Field Structure* (Springer-Verlag, Berlin Heidelberg, 1991).

[101] G. Birkenmeier, *Experimentelle Untersuchungen zur Struktur und Dynamik von Driftwellenturbulenz in Stellaratorgeometrie*, Doctoral thesis (Universität Stuttgart, 2012).

[102] K. Itoh, S.-I. Itoh, and A. Fukuyama, *Transport and Structural Formation in Plasmas* (CRC Press, London, 1999).

[103] M. Ramisch, *Scaling and Manipulation of Turbulent Structures in the Torsatron TJ-K*, Doctoral thesis (Universität Stuttgart, 2005).

[104] G. Wang *et al.*, Physics of Plasmas **20**, 092501 (2013).

[105] C. A. de Meijere *et al.*, Plasma Physics and Controlled Fusion **56**, 072001 (2014).

[106] N. Winsor, Physics of Fluids **11**, 2448 (1968).

[107] P. Angelino *et al.*, Physics of Plasmas **15**, 062306 (2008).

[108] T. Watari, Y. Hamada, A. Fujisawa, K. Toi, and K. Itoh, Physics of Plasmas **12**, 062304 (2005).

[109] V. B. Lebedev, P. N. Yushmanov, P. H. Diamond, S. V. Novakovskii, and A. I. Smolyakov, Physics of Plasmas **3**, 3023 (1996).

[110] S. V. Novakovskii, C. S. Liu, R. Z. Sagdeev, and M. N. Rosenbluth, Physics of Plasmas **4**, 4272 (1997).

[111] D. C. Leslie, *Developments in the theory of turbulence* (Clarendon Press, Oxford, 1973).

[112] E.-j. Kim, T. S. Hahm, and P. H. Diamond, Physics of Plasmas **8**, 3576 (2001).

182

[113] B. Scott, New Journal of Physics **7**, 92 (2005).

[114] L. Chen, Z. Lin, and R. White, Physics of Plasmas **7**, 3129 (2000).

[115] Ö. D. Gürcan, Physical Review Letters **109**, 155006 (2012).

[116] R. R. Trieling, O. U. V. Fuentes, and G. J. F. van Heijst, Physics of Fluids **17**, 087103 (2005).

[117] T. H. Stix, *Waves in Plasmas* (AIP-Press, New York, 1992).

[118] J. A. Krommes, Physics Reports **360**, 1 (2002).

[119] Y. Kodama and T. Taniuti, Journal of the Physical Society of Japan **47**, 1706 (1979).

[120] J. Boussinesq, Mémoires présentés par divers savants à l'Académie des Sciences Paris **23**, 1 (1877).

[121] V. Volterra, Nature **118**, 558 (1926).

[122] V. Berionni and Ö. D. Gürcan, Physics of Plasmas **18**, 112301 (2011).

[123] A. Yoshizawa, S.-I. Itoh, and K. Itoh, Plasma Physics and Controlled Fusion **45**, 321 (2003).

[124] M. A. Malkov, P. H. Diamond, and A. Smolyakov, Physics of Plasmas **8**, 1553 (2001).

[125] A. M. Dimits, T. J. Williams, J. A. Byers, and B. I. Cohen, Physical Review Letters **77**, 71 (1996).

[126] A. M. Dimits *et al.*, Physics of Plasmas **7**, 969 (2000).

[127] P. H. Diamond, Y.-M. Liang, B. A. Carreras, and P. W. Terry, Physical Review Letters **72**, 2565 (1994).

[128] M. a. Malkov, P. H. Diamond, and M. N. Rosenbluth, Physics of Plasmas **8**, 5073 (2001).

[129] P. Helander and D. J. Sigmar, *Collisional Transport in Magnetized Plasmas* (Cambridge University Press, Cambridge, 2005).

[130] Y. Idomura, M. Wakatani, and S. Tokuda, Physics of Plasmas **7**, 3551 (2000).

[131] B. Scott, Physics Letters A **320**, 53 (2003).

[132] B. Scott, Physics of Plasmas **8**, 447 (2001).

[133] U. Frisch, *Turbulence: The Legacy of A. N. Kolmogorov* (Cambridge University Press, Cambridge, 1995).

[134] J. Mathieu and J. Scott, *An Introduction to Turbulent Flow* (Cambridge University Press, Cambridge, 2000).

[135] K. Reuther, *The dependence of intermittent density fluctuations on collisionality in TJ-K*, Master's thesis (Universität Stuttgart, 2016).

[136] B. Pope, *Turbulent Flows* (Cambridge University Press, Cambridge, 2000).

[137] T. Huld, A. H. Nielsen, H. L. Pécseli, and J. Juul Rasmussen, Physics of Fluids B: Plasma Physics **3**, 1609 (1991).

[138] O. Grulke, F. Greiner, T. Klinger, and A. Piel, Plasma Physics and Controlled Fusion **43**, 525 (2001).

[139] C. Torrence and G. P. Compo, Bulletin of the American Meteorological Society **79**, 61 (1998).

[140] B. P. van Milligen, C. Hidalgo, and E. Sánchez, Physical Review Letters **74**, 395 (1995).

[141] E. Sanchez, T. Estrada, C. Hidalgo, B. Brañas, B. Carreras, and L. Garcia, Physics of Plasmas **2**, 3017 (1995).

[142] J. C. van den Berg, *Wavelets in Physics* (Cambridge University Press, Cambridge, 2004).

[143] M. Ramisch, *Bispektralanalyse turbulenter Fluktuationen am Experiment Teddi*, Diploma thesis (Christian-Albrechts-Universität zu Kiel, 2001).

[144] Y. C. Kim and E. J. Powers, IEEE Transactions on Plasma Science **7**, 120 (1979).

[145] M. J. Hinich and C. S. Clay, Reviews of Geophysics **6**, 347 (1968).

[146] H. L. Pécseli and J. Trulsen, Plasma Physics and Controlled Fusion **35**, 1701 (1993).

[147] C. Ritz and E. Powers, Physica D: Nonlinear Phenomena **20**, 320 (1986).

[148] P. Manz, *Strukturentstehung in Driftwellenturbulenz toroidaler Plasmen*, Doctoral thesis (Universität Stuttgart, 2009).

[149] C. P. Ritz, E. J. Powers, and R. D. Bengtson, Physics of Fluids B: Plasma Physics **1**, 153 (1989).

[150] D. Biskamp, *Magnetohydrodynamic Turbulence* (Cambridge University Press, Cambridge, 2003).

[151] M. D. Millionshchikov, Doklady Akademiia Nauk SSSR **22**, 236 (1941).

[152] M. D. Millionshchikov, Izvestiya Akademiia Nauk SSSR, Seriya Geograficheskaya i Geofizicheskaya **5**, 433 (1941).

[153] J. S. Kim, R. D. Durst, R. J. Fonck, E. Fernandez, A. Ware, and P. W. Terry, Physics of Plasmas **3**, 3998 (1996).

[154] H. Johnsen, H. L. Pécseli, and J. Trulsen, Physics of Fluids **30**, 2239 (1987).

[155] A. V. Filippas, R. D. Bengston, G.-X. Li, M. Meier, C. P. Ritz, and E. J. Powers, Physics of Plasmas **2**, 839 (1995).

[156] T. Windisch, O. Grulke, V. Naulin, and T. Klinger, Plasma Physics and Controlled Fusion **53**, 124036 (2011).

[157] C. Lechte, S. Niedner, and U. Stroth, New Journal of Physics **4**, 334 (2002).

[158] U. Stroth, F. Greiner, C. Lechte, N. Mahdizadeh, K. Rahbarnia, and M. Ramisch, Physics of Plasmas **11**, 2558 (2004).

[159] M. Ramisch, N. Mahdizadeh, U. Stroth, F. Greiner, C. Lechte, and K. Rahbarnia, Physics of Plasmas **12**, 032504 (2005).

[160] P. Manz, M. Ramisch, and U. Stroth, Physics of Plasmas **16**, 042309 (2009).

[161] M. Ramisch, P. Manz, U. Stroth, G. Birkenmeier, S. Enge, E. Holzhauer, A. Köhn, and B. Nold, Plasma Physics and Controlled Fusion **52**, 124015 (2010).

[162] M. Ramisch *et al.*, Contributions to Plasma Physics **50**, 718 (2010).

[163] T. Ullmann, *Untersuchung zur Abhängigkeit zonaler Strömungen von Hintergrundplasmaströmungen*, Staatsexamen thesis (Universität Stuttgart, 2015).

[164] A. A. Peshwaz, *Magnetic Configuration Effects on the Torsatron TJ-K Plasma Parameters and Turbulent Transport*, Master's thesis (Universität Stuttgart, 2009).

[165] Z. Ivady, *Influence of magnetic islands on transport in a toroidal plasma*, Diploma thesis (Universität Stuttgart, 2011).

[166] T. Happel, *Influence of Limiters on Plasma Equilibrium and Dynamics in the Torsatron TJ-K*, Diploma thesis (Christian-Albrechts-Universität zu Kiel, 2005).

[167] B. Schmid, *Bildung turbulenter Strukturen im Randbereich magnetisch eingeschlossener Plasmen*, Diploma thesis (Universität Stuttgart, 2011).

Bibliography

[168] MCC, written in IDL/C/C++ by M. Ramisch (2005).

[169] H. Höhnle, *Frequenzgesteuerte Arrayantenne zur Elektron-Zyklotron-Resonanz- Heizung mit Bernstein-Wellen am Torsatron TJ-K*, Diploma thesis (Universität Stuttgart, 2008).

[170] A. Köhn, *Mikrowellenheizung von überdichten Plasmen in TJ-K*, Diploma thesis (Christian-Albrechts-Universität zu Kiel, 2005).

[171] G. Birkenmeier, *Experiments and modeling of transport processes in toroidal plasmas*, Diploma thesis (Universität Stuttgart, 2008).

[172] A. Köhn, *Investigation of microwave heating scenarios in the magnetically confined low-temperature plasma of the stellarator TJ-K*, Doctoral thesis (Universität Stuttgart, 2010).

[173] C. Gourdon, *Programme optimise de calculs numerique dans les configurations magnetique toroidales* (CEN, Fontenay aux Roses, 1970).

[174] G. Grieger *et al.*, in *Plasma Physics and Controlled Fusion Research (Proc. 13th Int. Conf., Washington, DC, 1990), Vol. 3* (IAEA, Vienna, 1991), pp. 525–532.

[175] G. Grieger *et al.*, Physics of Fluids B: Plasma Physics **4**, 2081 (1992).

[176] A. H. Boozer, Physics of Fluids **24**, 1999 (1981).

[177] S. Hamada, Nuclear Fusion **2**, 23 (1962).

[178] S. Niedner, *Numerical studies of plasma turbulence for comparison with measurements at TJ-K*, Doctoral thesis (Christian-Albrechts-Universität zu Kiel, 2002).

[179] Z. Huang, *Probe Measurement of Electron Temperature Dynamics in TJ-K and ASDEX Upgrade*, Master's thesis (Universität Stuttgart, 2011).

[180] B. Nold, *Untersuchung turbulenter Strukturen am Rand magnetisierter Plasmen*, Doctoral thesis (Universität Stuttgart, 2012).

[181] S. Enge, G. Birkenmeier, P. Manz, M. Ramisch, and U. Stroth, Physical Review Letters **105**, 175004 (2010).

[182] G. Birkenmeier, H. Höhnle, A. Köhn, M. Ramisch, and U. Stroth, IEEE Transactions on Plasma Science **36**, 1092 (2008).

[183] A. Köhn, G. Birkenmeier, E. Holzhauer, M. Ramisch, and U. Stroth, Plasma Physics and Controlled Fusion **52**, 035003 (2010).

[184] A. Köhn, Á. Cappa, E. Holzhauer, F. Castejón, Á. Fernández, and U. Stroth, Plasma Physics and Controlled Fusion **50**, 085018 (2008).

[185] C. Lechte, J. Stöber, and U. Stroth, Physics of Plasmas **9**, 2839 (2002).

[186] E. Häberle, *Skalierung turbulenter Strukturen im Torsatron TJ-K*, Diploma thesis (Universität Stuttgart, 2007).

[187] N. Mahdizadeh, *Investigation of Three-Dimensional Turbulent Structures in the Torsatron TJ-K*, Doctoral thesis (Universität Stuttgart, 2007).

[188] R. J. Fonck *et al.*, Plasma Physics and Controlled Fusion **34**, 1993 (1992).

[189] L. Giannone *et al.*, Physics of Plasmas **1**, 3614 (1994).

[190] U. Stroth, Plasma Physics and Controlled Fusion **40**, 9 (1998).

[191] P. C. Liewer, Nuclear Fusion **25**, 543 (1985).

[192] J. C. Mcwilliams, Journal of Fluid Mechanics **146**, 21 (1984).

[193] R. Benzi, G. Paladin, S. Patarnello, P. Santangelo, and A. Vulpiani, Journal of Physics A: Mathematical and General **19**, 3771 (1986).

[194] T. Dubos, A. Babiano, J. Paret, and P. Tabeling, Physical Review E **64**, 036302 (2001).

[195] P. Tabeling, Physics Reports **362**, 1 (2002).

[196] L. Cui, G. R. Tynan, P. H. Diamond, S. C. Thakur, and C. Brandt, Physics of Plasmas **22**, 050704 (2015).

[197] L. Cui, A. Ashourvan, S. C. Thakur, R. Hong, P. H. Diamond, and G. R. Tynan, Physics of Plasmas **23**, 055704 (2016).

[198] M. Ramisch, F. Greiner, N. Mahdizadeh, K. Rahbarnia, and U. Stroth, Plasma Physics and Controlled Fusion **49**, 777 (2007).

[199] T. Herzog, *Charakterisierung des turbulenten Transports in der Abschälschicht limitierter Plasmen*, Bachelor's thesis (Universität Stuttgart, 2012).

[200] G. Fuchert, *Dynamics and structure analysis of coherent turbulent structures at the boundary of toroidally confined plasmas*, Doctoral thesis (Universität Stuttgart, 2013).

[201] O. Grulke, *Investigation of large-scale spatiotemporal fluctuation structures in magnetized plasmas*, Doctoral thesis (Christian-Albrechts-Universität zu Kiel, 2001).

[202] B. Schmid, P. Manz, M. Ramisch, and U. Stroth, Physical Review Letters **118**, 055001 (2017).

[203] S. E. Parker, Y. Chen, W. Wan, B. I. Cohen, and W. M. Nevins, Physics of Plasmas **11**, 2594 (2004).

[204] B. Schmid, P. Manz, M. Ramisch, and U. Stroth, New Journal of Physics **19**, 055003 (2017).

[205] Z. Yan, J. H. Yu, C. Holland, M. Xu, S. H. Müller, and G. R. Tynan, Physics of Plasmas **15**, 092309 (2008).

[206] C. Holland, J. H. Yu, A. James, D. Nishijima, M. Shimada, N. Taheri, and G. R. Tynan, Physical Review Letters **96**, 195002 (2006).

[207] G. R. Tynan, C. Holland, J. H. Yu, A. James, D. Nishijima, M. Shimada, and N. Taheri, Plasma Physics and Controlled Fusion **48**, S51 (2006).

[208] G. Birkenmeier, M. Ramisch, P. Manz, B. Nold, and U. Stroth, Physical Review Letters **107**, 025001 (2011).

[209] A. Bhattacharjee, J. E. Sedlak, P. L. Similon, M. N. Rosenbluth, and D. W. Ross, Physics of Fluids **26**, 880 (1983).

[210] R. E. Waltz and A. H. Boozer, Physics of Fluids B: Plasma Physics **5**, 2201 (1993).

[211] P. Manz *et al.*, in *Proc. of the 44th EPS Conference on Plasma Physics, Belfast, Northern Ireland* (The European Physical Society, Belfast, 2017), pp. 1–4.

[212] M. Ramisch, G. Birkenmeier, A. Köhn, and U. Stroth, in *Proc. of the 38th EPS Conference on Plasma Physics, Strasbourg, France* (The European Physical Society, Strasbourg, 2011), pp. 1–4.

[213] M. Nadeem, T. Rafiq, and M. Persson, Physics of Plasmas **8**, 4375 (2001).

[214] S. Futatani, S. Benkadda, Y. Nakamura, and K. Kondo, Physics of Plasmas **15**, 072506 (2008).

[215] P. D. Dura, B. Hnat, J. Robinson, and R. O. Dendy, Physics of Plasmas **19**, 092301 (2012).

[216] V. Naulin, A. Kendl, O. E. Garcia, A. H. Nielsen, and J. J. Rasmussen, Physics of Plasmas **12**, 052515 (2005).

[217] A. Kendl and B. D. Scott, Physics of Plasmas **12**, 064506 (2005).

[218] S. B. Korsholm, P. K. Michelsen, V. Naulin, J. J. Rasmussen, L. Garcia, B. A. Carreras, and V. E. Lynch, Plasma Physics and Controlled Fusion **43**, 1377 (2001).

[219] P. Manz, M. Xu, S. C. Thakur, and G. R. Tynan, Plasma Physics and Controlled Fusion **53**, 095001 (2011).

[220] K. M. Weber, *Investigation of equilibrium flows in a toroidal plasma*, Diploma thesis (Universität Stuttgart, 2009).

[221] H.-P. Kriegel, P. Kröger, and A. Zimek, ACM Transactions on Knowledge Discovery from Data **3**, 1 (2009).

[222] H.-P. Kriegel, P. Kröger, J. Sander, and A. Zimek, Wiley Interdisciplinary Reviews: Data Mining and Knowledge Discovery **1**, 231 (2011).

Danksagung

An dieser Stelle möchte ich mich bei allen bedanken, die zum Gelingen dieser Arbeit beigetragen haben, insbesondere bei

Herrn Prof. Dr. U. Stroth für den Hauptbericht, die gute Betreuung und die Möglichkeit und Freiheit diese Doktorarbeit anfertigen zu dürfen,

Herrn Prof. Dr. J. Starflinger für die Übernahme des Mitberichts,

Herrn Prof. Dr. E. Laurien für den Prüfungsvorsitz,

Herrn Prof. Dr. T. Hirth und Herrn Prof. Dr. G. Tovar für die freundliche Aufnahme am Institut,

Dr. Mirko Ramisch und Dr. Peter Manz für die kompetente und geduldige Betreuung, die vielen Ideen und die Hilfe, ohne die diese Arbeit so nicht möglich gewesen wäre,

Stefan Wolf für die freundliche Arbeitsatmosphäre und dafür, dass er bei einem Problem sofort weiterhilft,

Dr. Alf Köhn, Til Ullmann, Gabriel Sichardt, Stephen Garland, Bernhard Roth und Dr. Eberhard Holzhauer für die Hilfe beim Messen und die vielen Diskussionen,

Frau Wagner, die immer überaus freundlich und hilfsbereit war,

dem TJ-K Team und den Gruppen MT und PT,

meinen Eltern, die mich immer unterstützt haben und ohne die ich bis hierhin nicht gekommen wäre,

und Claudia, die immer für mich da war.

www.ingramcontent.com/pod-product-compliance
Lightning Source LLC
Chambersburg PA
CBHW060305220326
41598CB00027B/4240